MOTORCYCLING AND LEISURE

Human Factors in Road and Rail Transport

Series Editors

Today's society must confront major land transport problems. The human and financial costs of vehicle accidents are increasing, with road traffic accidents predicted to become the third largest cause of death and injury across the world by 2020. Several social trends pose threats to safety, including increasing car ownership and traffic congestion, the increased complexity of the human-vehicle interface, the ageing of populations in the developed world, and a possible influx of young vehicle operators in the developing world.

Ashgate's 'Human Factors in Road and Rail Transport' series aims to make a timely contribution to these issues by focusing on the driver as a contributing causal agent in road and rail accidents. The series seeks to reflect the increasing demand for safe, efficient and economical land-based transport by reporting on the state-of-theart science that may be applied to reduce vehicle collisions, improve the usability of vehicles and enhance the operator's wellbeing and satisfaction. It will do so by disseminating new theoretical and empirical research from specialists in the behavioural and allied disciplines, including traffic psychology, human factors and ergonomics.

The series captures topics such as driver behaviour, driver training, in-vehicle technology, driver health and driver assessment. Specially commissioned works from internationally recognised experts in the field will provide authoritative accounts of the leading approaches to this significant real-world problem.

Motorcycling and Leisure

Understanding the Recreational PTW Rider

PAUL BROUGHTON
Owl Research Ltd, UK

&

LINDA WALKER
University of Stirling, UK

CRC Press is an imprint of the
Taylor & Francis Group, an **informa** business

CRC Press
Taylor & Francis Group
6000 Broken Sound Parkway NW, Suite 300
Boca Raton, FL 33487-2742

First issued in paperback 2019

ISBN-13: 978-0-7546-7501-3 (hbk)
ISBN-13: 978-0-367-38560-6 (pbk)

Visit the Taylor & Francis Web site at
http://www.taylorandfrancis.com

and the CRC Press Web site at
http://www.crcpress.com

Contents

List of Figures

List of Tables

Forewords

It is a strange paradox that the group of road users most at risk from serious or fatal injury – motorised two-wheelers – is the least investigated by safety researchers. Perhaps the popular stereotype of a Hell's Angel, a 'weekend warrior in one piece leathers', is not viewed as a particularly accessible participant to a young researcher armed with an attitudinal questionnaire; perhaps the prevailing culture embraces a tacit, prejudiced view of the motorcyclist along the lines 'you've chosen an unstable, high risk form of transport, mate: you take the consequences', a view captured in Leonard Evans' pithy remark: 'Buy your son a motorbike for his last birthday' (Evans 1991). But the fact is that of the 1.6 million motorcyclists in the UK, half are over 40 and are in managerial and professional jobs.

In this unique and highly readable book, you will learn about who rides, what they ride and why they do it. The extensive coverage includes basic bike controls, licensing, training and testing, the nature and causes of crashes, how to ride safely and the many pleasures (and displeasures) of riding. Not only will it be a 'must read' for every person who has ever got up on a powered two-wheeler – or is thinking about it – but it should be read by every car and bus and truck driver for its insights into what the biker's task is like and how drivers can cooperate to make life on the road more pleasant – and safer – for both.

The authors use the latest theory and research to take you into the minds of motorcyclists in a comprehensive way. Written with obvious deep knowledge and considerable passion for the two-wheeler, here is the heart and soul of motorcycling.

Ray Fuller
Trinity College Dublin

Bikers: who are these people? With a post-war public image as hedonistic individualist outlaws bent on thrill- and adventure-seeking derived from Marlon Brando and Lee Marvin in the iconic 50s film *The Wild One*; from the drug- and alcohol-fuelled violent exploits of Sonny Barger and the Oakland Chapter of the Hell's Angels as vividly documented by Hunter S. Thompson; from Peter Fonda, Dennis Hopper and Jack Nicholson in *Easy Rider* as they 'head out on the highway, looking for adventure' to the soundtrack of Steppenwolf's *Born to be Wild*; and the 60s seaside confrontations at Margate and Brighton of UK mods and rockers, how come these thrill-seekers defer gratification for so long before each run out, taking ages creaking themselves into their long leathers and fiddling with their sprockets?

In this up-to-date and timely volume Paul Broughton and Linda Walker dismantle the myths about motorcycling and replace them with a much more illuminating data- and theory-based account of the pleasures and perils of modern biking, placing enjoyment, not risk taking, at the heart of the relationship between the recreational rider and his (and occasionally her) machine. In searching for enjoyment from riding, riders seek the space between boredom and anxiety, and in seeking challenge they are warriors, but not outlaws.

Theories of behaviour should precede, and thereby inform, theories of behaviour change. This volume provides an excellent antidote to the myths and in advancing our understanding of recreational riding behaviour offers clear pointers to how to reduce the carnage that still results from the inherent vulnerability of Powered Two-Wheeler riders when their trajectories intersect with a dry stone wall on a bend or an unobservant motorist at a junction.

Professor Steve Stradling
Transport Research Institute
Napier University

Preface

Motorcyclists are rough, dangerous, risk seeking psychopaths – or at least that is the impression often given; however, the authors did not recognise this description with respect to those motorcyclists that they knew. It was this that started the passion into researching who these riders were and 'what made them tick'. The exploration of rider behaviour demonstrated that riders are not these extreme risk seekers, rather that these ride for the love of biking. One motorcyclist explained his passion this way, 'Motorcycling, anything else is just transport'.

Dr Paul S. Broughton
Dr Linda Walker

Acknowledgements

We are indebted to the following colleagues for their advice, assistance and support: Professor Steve Stradling, Professor Ray Fuller and Dr Cris Burgess. Thanks also to other colleagues and friends who have sparked ideas and inspired this book.

This book would not have been possible without the many riders who gave up their time to share their experiences, special thanks goes to the members of the UK Bikers Forum and the Scottish Motorcycle Club. Also, we are grateful for the wisdom imparted from Zomb!e, who will be sadly missed.

We would also like to acknowledge the help and support of the Motorcycle Industry Association for their willingness to share information for the common good of motorcycle safety.

This book could not have been possible without support of friends, family and childminders.

Abbreviations

2DLD	Second Driving Licence Directive
3DLD	Third Driving Licence Directive
ABS	Advanced Braking System
ACEM	Association des Constructeurs Européens de Motocycles
AISS	Arnett's Inventory of Sensation Seeking
Bhp	Brake horse power
BMF	British Motorcycle Federation
CBE	Challenge Based Enjoyment
CBT	Compulsory Basic Training
DAS	Direct Access Scheme
DfT	Department for Transport
DSA	Driving Standards Agency
DVLA	Driver Vehicle Licensing Authority
EC	European Community
EU	European Union
FEMA	Federation of European Motorcyclists' Associations
IAM	Institute of Advanced Motorists
Kmh	Kilometres per hour
KSI	Killed or Seriously Injured
Kw	Kilowatt
MAG	Motorcycle Action Group
MAIDS	Motorcycle Accident In Depth Study
MCIA	Motorcycle Industry Association
Mph	Miles per hour
PI	Performance Index
PTW	Powered Two-Wheeler
RBE	Rush Based Enjoyment
RoSPA	Royal Society for the Prevention Accidents
SHARP	Safety Helmet and Assessment Rating Programme
SMIDSY	'Sorry Mate I Didn't See You'
UK	United Kingdom
VAT	Value Added Tax

This book is dedicated to Beavis

Chapter 1
An Introduction to Motorcycles

This book covers some of the latest research on Powered Two-Wheelers (PTWs) dealing with issues such as risk, who rides PTWs, why they ride, how riders compare to car drivers and the road safety implications for those who choose this mode of transport.

A PTW, as the name suggests, is a vehicle that has two wheels and is propelled by a power unit. This therefore includes motorcycles, mopeds and scooters but would exclude pushbikes. The power unit does not have to be combustion engine based so a battery-powered machine would also fit into the PTW category, however, the majority of PTWs are powered by petrol-based combustion engines. This chapter offers a brief introduction to the world of the PTW with a brief outline of issues and terminology associated with PTWs and their users.

PTW Usage

There has been a steady rise in motorcycle ownership over recent years (RoSPA 2001). In 2006 there were 133,077 new bikes registered in the UK (MCIA 2007b) and according to Mintel (2004), even though the market is slowing down from the rapid growth it experienced at the end of the twentieth century, it still remains buoyant. Around 2.3 per cent of households own at least one PTW with PTW use accounting for about 1 per cent of total annual road mileage (DfT 2006a).

Mintel predict that leisure biking will be a market driver with people turning to biking as a way to relieve the stress of work but the motorcycle industry will have to compete with other leisure sectors, especially those which offer a safer alternative to stress relief. Chorlton and Jamson (2003) showed that there is a shift in the nature of motorcycling, with more machines now being purchased for leisure riding rather than for functional journeys. They suggest that those who ride mainly for leisure generally have larger capacity bikes and are mainly long-term or returning riders; leisure riding is a key reason for PTW use (Chorlton and Jamson 2003).

Possibly because motorcycle usage is seen as being primarily a leisure activity, the role of the motorcycle as a viable means of transport has sometimes been trivialised, yet within Europe they account for around 3 per cent of surface transport; Europe's railways only account for 6 per cent (Diekmann 1996). Motorcycles can offer a cheap and energy efficient means of transport, giving options for some who do not have access to a car, as well as offering a valuable alternative transport method to car owners. Given the specific transport needs that motorcycles satisfy, it would be difficult to replace them whether for commuting or for leisure. Since the Diekmann Report more Europeans have begun to use PTWs as an urban mode of transport, a method that preserves their freedom of mobility and helps them to get through traffic congestion (ACEM 2000).

If motorcycling was made easier, safer and more convenient then more people may ride which would in turn reduce congestion as well as improving the environment (ACEM 2000). A 2004 Federation of European Motorcyclists' Associations (FEMA) report also supports this view as well as expressing that the riding of a PTW is a meaningful leisure activity that improves the quality of life for millions of European citizens; the average age of the European rider is rising and more women are now riding (FEMA 2004).

In the UK there is some evidence that a considerable amount of 'biking' is carried out for commuting purposes (RoSPA 2001). A survey of participants of the Bikesafe scheme in Scotland, an initiative run by police forces in the United Kingdom to help to lower the number of motorcycle rider casualties, showed that 93 per cent of respondents used their bikes for pleasure and 51 per cent used their bikes for 'getting to work' (Ormston, Dudleston, Pearson and Stradling 2003). The Department for Transport (DfT) reported that in terms of distance, 56.3 per cent of all trips are for work, business or education, with 20.7 per cent for leisure (DfT 2004a), however, the figures from the 2006 national travel survey suggests that 42.4 per cent of trips are for leisure (DfT 2006c). The average miles ridden per week is 88.8 miles taking an average of 3.4 hours, while for car drivers the weekly average mileage is around 70 miles taking an average of 3.0 hours (DfT 2006c).

While there are advantages to using PTWs there are also disadvantages. Riding a motorcycle is a more complex operation than that of driving a car (Mannering and Grodsky 1995) and this, coupled with the additional vulnerability of PTW users (RoSPA 2001), gives rise to the perception that the risk of riders is higher than that of other motorised vehicles. The statistics show that the risk of having a crash is not higher for PTW users (ACEM 2004), but that the risk of serious or fatal injury is (DfT 2006a).

The Risk and the Rider

This higher risk of serious injury or even death suggests that those who ride must be consummate risk takers. Often the view of those who ride PTWs is that of a young, risk seeking outlaw, covered in grease unlikely to have not washed for quite some time. A closer examination though shows that riders often work in managerial positions and use their bike as a second mode of transport. This dichotomy of image against reality is a subject worthy of further investigation. An examination of riders, their profiles and reason for using PTWs seeks to explore some of the paradoxes of biker image with real world PTW use.

Motorcycles are different to cars; with one the major differences between the two vehicle modes being the vulnerability of the PTW rider. This vulnerability, caused in part by the exposed nature of the rider, gives rise to the perception of PTW riding as a high risk activity.

Safety is an issue that is intrinsically linked with the riding of a PTW and with the increase in bike use has come a rise in casualties from motorcycle-related incidents. In Great Britain, Killed or Seriously Injured (KSI) crashes rose from 5,717 in 1996 to 6,255 in 2004 (DfT 2006a), a rise of nearly 10 per cent at a time when there has been a reduction for other road users (DfT 2004c). Although this percentage rise is less than the percentage increase in bike ownership for the same period (739,000 to 1,191,000 – 61 per cent), it is still a major concern.

Types of Powered Two-Wheelers

Powered Two-Wheelers cover a broad spectrum of vehicles. They range from small scooters, ideal for short journeys and running around town, to extremely powerful machines capable of speeds well in excess of the legal speed limits. There are a number of ways that PTWs can be classified but for the purpose of this book six groupings are used: sportsbike; tourer; sports tourer/all rounder; classic/custom/cruiser; off-road/trail/adventure tourer; and moped/scooter.

The Sportsbike

The sportsbike is essentially a consumer version of a race track bike and tends to be a lightweight, high-powered, very fast machine with the rider position leant forward to minimise wind resistance.

Within the sportsbike category there are other sub-types. The ‘Pocket Rocket’ is a small sportsbike with a relatively small engine, rarely above 400 cc, and is the ‘little brother’ to the superbike and hyperbike. The superbike title mainly is given to race replica machines that have a capacity of more than 750 cc, while hyperbikes are large, very fast sportsbikes often capable of speeds in excess of 170 mph. With hyperbike development pushing the top speeds towards the 200 mph mark common sense prevailed and there is now a voluntary agreement amongst manufacturers to limit the top speed of these bikes to 300 km/h (186 mph). The engines used within sportsbikes are often designed to be high revving, with useable engine speeds of 9,000 revolutions per minute (rpm) not being unusual. Most cars’ useable range is 2,000 rpm to 4,000 rpm with a maximum of around 6,500 rpm.

The Tourer

The purpose of the tourer is for comfortable high mileage riding and it is therefore designed around rider comfort. The rider is seated in an upright position and the bike has the capacity to carry large amounts of luggage. The engine of tourer bikes is designed with a high torque for long journeys carrying heavy loads. These bikes tend to be heavier than sportsbikes.

Figure 1.1 MV Agusta 750s (courtesy of www.bikemotel.nl)

Figure 1.2 BMW K100 RT

Sportstourer/All Rounder

The sportstourer, or 'all rounder', is exactly what it sounds like: a cross between a sportsbike and a tourer with the blend of attributes allowing long distance riding at higher speeds. This compromise often means that the rider is not as comfortable as they would be on a tourer and the machine is not capable of the very high speeds of the pure sportsbikes.

Classic/Custom/Cruiser

These machines are often in the style of American bikes from the 1930s through to the 1960s. These are normally big engine bikes with the rider sat in an upright relaxed position and the feet in a more forward placement than on other PTWs. These bikes often have no fairing and lots of chrome with Harley Davidson being a name that is synonymous with the genre of bike.

Figure 1.3 Kawasaki GPZ 500S

Figure 1.4 Harley Davidson (courtesy of Andrew Bunker)

Off-Road/Trail/Adventure Tourer

Off-road machines are designed for riding through rough, muddy and uneven countryside. They have bulky tyres designed to get grip on muddy surfaces and suspension systems designed to compensate for rough terrain. These machines are not geared for top end speed but to deliver power in a suitable manner for rough ground riding.

The adventure tourer is a sub-class within these off-road type machines that is designed to allow riders to tour on a bike that has some off-road capabilities. The adventure tourer end of the market has been brought into focus by the recent televised trips made by Ewan McGregor and Charley Boorman for the BBC. For these two programmes, '*The Long Way Round*' and '*The Long Way Down*', the BMW R1150GS Adventure was the bike of choice.

Moped/Scooter

Mopeds and scooters are normally at the lower end of the engine capacity, but there are now scooters that have significantly larger engine sizes accompanied by a higher power output than a few years ago, for example, the Suzuki Burgman 650 is a 650 cc scooter that is capable of speeds in excess of 110 mph.

Mopeds have an engine size of less than 50 cc, and must have a maximum speed of no more than 50 km/h (31 mph). From June 2003 all EC Type Approved mopeds have a maximum speed of 45 km/h (28 mph).

Figure 1.5 One of the BMW R1150GS Adventures used for '*The Long Way Round*'

Figure 1.6 FJS 600 ABS Honda Silver Wing (courtesy of Pete McBride)

Other Bike Types

Some may argue that there are other bike types, but machines that they purport to be classified outwith the above mostly fit into one of these categories. For example, the naked bike is usually a stripped down version of another bike with the fairing and other cosmetic parts removed.

Performance Index

Engine size is often used in categorising motorcycles for analysing PTW crashes (for example, see DfT 2004c; DfT 2006a; EuroRap 2004; Huang and Preston 2004; Sexton, Hamilton, Baughan, Stradling and Broughton 2006; Yannis, Golias, and Papadimitriou 2005). However, this method of categorising machines does not take into account the actual performance capability of the machine. For example, a Honda SL650 has an engine size of 649 cc, but only a top speed of 95 mph while a Suzuki GSXR 600 has a smaller engine size of 600 cc yet a top speed of 160 mph (UBG 2006). This prompted Broughton (2007) to use the bike power and weight data to develop a performance index and thus allow a better comparison of the bikes. The equation for the performance index (P_i) is:

$$P_i = \frac{Power}{Weight} \times TopSpeed$$

Where the relevant data exists, nominally the make and model of the motorcycle, then the performance index (P_i) can be used to categorise bike data in a meaningful way.

Riders may choose their PTW based on performance, functionality or image. Some may choose different types of bikes depending on their circumstances while for others they have a loyalty to a particular brand, make or even a particular model. Although there are a variety of bike types, the fundamentals of riding PTWs are similar for all. However, riding a PTW is substantially different to driving a car and therefore to aid the reader in gaining a better understanding of bike use, some basic riding skills are now described.

Basics of Riding a Powered Two Wheeler

Although PTWs and cars are governed on British roads with the same generic set of legislation, albeit with some specific clauses such as riders having to wear a helmet (DirectGov 2008), the riding of a motorbike is very different from driving a car, varying in such aspects as the position of the controls and the skills required.

Bike Controls

No matter what vehicle you are driving it is important that the controls are positioned for easy use. For a PTW this produces its own set of problems, for example, within a car the driver would rarely be wearing safety equipment that would restrict their movements or tactile capabilities but riders are restricted by their safety equipment such as gloves.

The description of the controls here are true for most machines, however, as with most things there are a few PTWs that deviate from this norm. On a PTW the majority of the controls are situated on or near the end of the handlebars with the switches designed to allow gloved operation. The main exceptions to this placement are the gear change and rear brake, which are foot controls.

The majority of PTWs have independent front and rear brakes; the front brake is hand-operated by a lever on the right hand side of the handlebars and the rear brake is operated by a lever activated by the right foot. However, more bikes are now available with linked brakes, allowing braking on both wheels to be controlled via one lever, for example, the Honda CBR1100XX Blackbird.

The clutch, in the form of a pull lever, is situated on the left hand side of the handlebar. The other controls on the left hand side are positioned so that the left thumb can operate them and include the horn, hi-beam light control, passing lights and indicator control. The normal indicator control on a PTW differs from those found on a car in two significant ways: indicators do not auto-cancel, so a rider must always be sure that they are turned off after use so as not to send false intentions to other road users; an indicator is turned on by moving the button to the left or right and is cancelled by pushing the button in.

The controls found on the right hand side of the handlebars include the pull lever that operates the front brake and the throttle control. The throttle operates on a twist-grip principle – that is the handgrip is twisted to alter the amount of throttle being applied to the engine, clockwise (towards the rider) for more throttle and anti-clockwise (away from the rider) for less throttle. The button used to start the engine is situated on the right hand side of the handlebars along with the emergency engine stop switch that can be used to kill the engine quickly if needed. The on/off controls for the lights are also placed on this side of the handlebars.

The left foot is used to change gears by pushing the foot-operated lever either up or down. Motorcycles have a sequential gear box so each gear must be selected in turn and gears cannot be missed; for example, to get from second to fourth, third gear must be selected. Generally pushing the level down will select a lower gear and pushing it upwards will select a higher ratio. Neutral is situated between first and second so to place the bike into first from neutral the gear lever is pushed down, then to move into second the lever is moved upwards, and then up again to change into third, hence a six speed gearbox will often be referred to as one down five up. The right foot is used to work the rear brake by depressing it; the harder it is depressed the harder the brake is applied.

Basic Bike Control

As with driving a car, the knowledge of where the controls are situated and what each control does is not enough to allow the safe operation of the vehicle. In the UK, basic PTW riding skills are taught to a rider when they undertake compulsory basic training (CBT). The elements of the CBT are discussed later in this chapter. This training deals with some of the specific skills that are required for riding a PTW.

Even before the engine has started, some differences between PTW riding and car driving are apparent. When a PTW does not have a rider sat upon it, it has to be supported. Most machines have two methods of doing this, a side-stand and a centre-stand. The side-stand is a flip-down device that allows the bike to be supported by leaning the weight of the bike against it. A side-stand can be activated with the rider still on the bike by flipping the stand down and then leaning the bike onto the stand before getting off the bike. The centre-stand is a more substantial method of supporting the bike and this cannot be operated while the rider is on the bike. The centre-stand is in the middle of the bike and supports the bike in an upright position, often with the rear wheel off the ground. To place the PTW on to the centre-stand the rider must balance the bike while standing beside it and then lift the bike back and up onto the stand.

When a moving bike comes to rest the rider must support the bike by putting at least one foot onto the ground, and the bike must also be held on either the front or rear brake to prevent it from moving. Unlike a car, most bikes do not have a brake that can be activated and left on as in the case of the car's handbrake. For the majority of time when a rider is sitting on a stationary bike it should be held on the back brake by using the right foot with the weight of the machine supported using the rider's left leg.

Most modern PTWs use an electric starter operated by a push button situated on the right hand side of the handlebars. Older machines tended to have a 'kick start' system where the bike is started by the rider standing upon the kick start lever and pushing it downwards, which turns the engine over and starts the machine. Regardless of the starting system, prior to starting the bike the ignition key has to be switched to the on position, illuminating the warning lights on the display. The rider should check these to ensure the bike is in neutral before starting. For older bikes that do not have an electrical indication of being in neutral, the bike can be rocked back and forth to ensure that it is not engaged in any gear.

Once the bike has been started the rider can prepare to pull away. The order of events for doing this is the same as for driving a car; clutch in, select gear and increase the engine speed whilst balancing against the clutch so that it allows the vehicle to pull off. However, on a PTW there are some other complications. First, to select a gear the left foot must be used but as this is the foot that is supporting the machine the weight of the bike must be switched so that the right foot becomes the supporting foot. But as the right foot is being used to hold the machine still by using the rear brake, the front brake must first be activated by using the right hand. Therefore the basic procedure, in addition to normal observations, for pulling away on a bike is:

- apply the front brake;
- release the rear brake and put the right foot down;
- shift the weight of the machine onto the right leg and ensure it is balanced;
- bring the left foot onto the footrest;
- pull the clutch in using the left hand;
- select first gear by pressing the gear change downwards with the left foot. The neutral light on the display should go out;
- put the left foot back onto the ground;
- shift the weight of the PTW from the right to the left leg and balance the bike;
- put the right foot back onto its footrest;
- apply the rear brake using the right foot;
- release the front brake (right hand);
- increase the speed of the engine by operating the twist grip with the right hand;
- move the clutch to biting point by slowly releasing it with the left hand;
- balance the engine speed and clutch and release the back brake as the bike starts to want to pull away;
- as the machine starts to move off, place the left foot onto the footrest.

Once the bike is moving, further progress can be made by moving through the gears on the sequential gearbox. To change up a gear the clutch is operated using the left hand on the pull lever, at the same time engine speed is reduced by closing the throttle off by twisting the twist grip anti-clockwise with the right hand. The gear change is actuated by moving the gear lever upwards with the front part of the left foot, the clutch is then slowly released while being balanced against an increase in engine speed. Changing down a gear is very similar except that the gear changer is pressed downwards by the sole of the left foot instead of upwards. When the PTW comes to a stop it can be put into neutral by an upward half movement of the gear changer to select a gear between first and second, this of course assumes that the machine was in first gear after stopping. While putting the machine into neutral it must always be held on a brake, so the shuffle between the left and right supporting leg as well as changing from using the front/rear brakes would need to be carried out in a similar fashion to that used when pulling off.

When attempting to slow down a motorcycle, engine braking should be used where possible as use of the brakes can make the machine unstable. Engine braking is simply shutting off the throttle and allowing the engine to slow the vehicle up with more effective engine braking being achieved by selecting a lower gear. If the use of the brakes is required then these should be applied with a balance between the front and rear brake with the majority of the braking being carried out with the front brake. In dry conditions about 80 per cent of the braking effort should be the front brake with the ratio moving more towards 50/50 in wet conditions. With the nature of bikes being unstable braking needs to be carried out with caution and in a controlled way; it should only occur while the bike is travelling in a straight line and not leaning from the perpendicular otherwise the front wheel is liable to slide out from under the bike.

The above describes some of the mechanics of riding a PTW, but to ride safely a number of additional skills are also needed, such as road craft, good hazard perception and a high observational competence. While some of this is learnt and honed by experience, the basics are taught to riders during their passage to obtaining a licence.

How to Get a Motorcycle Licence in the United Kingdom

The route to a licence to ride a PTW in the UK that will allow a person to be qualified to ride any bike is not the same, or as simple, as for a car (DfT 2003b). This section gives a brief overview of the current situation with regard to gaining of a licence in the UK.

Current Licence Types

There are three types of full PTW licence that a rider can hold (DfT 2007d). The 'P' licence permits the rider to ride a moped without the restrictions of a learner, they do not have to display 'L-plates' and they can carry passengers. However, mopeds are not permitted on motorways. The 'A1' licence allows a rider to ride a machine with a capacity of up to 125 cc with a power output not exceeding 11 kW. The 'A' licence is more complex with two stages attached to it. On passing a practical test on a machine of around 125 cc the rider can use, without restrictions, a bike of up to 25 kW in power to weight ratio that does not exceed 0.16 kW/kg. These power restrictions are in force for two years after which the rider is entitled to ride a PTW of any size or power. Riders over the age of 21 can bypass the two-year probationary period by doing what is commonly referred to as DAS or Direct Access Scheme. To do this they must complete their practical riding test on a bike with a minimum power output of 35 kW. To enable a rider to practise for this test they are allowed to ride a machine that is acceptable for their test on the public road with 'L-plates' but they must be accompanied by an approved instructor who is on another motorcycle and can talk to the pupil via a one-way radio.

All new riders, regardless of which class of licence sought, have to complete compulsory basic training (CBT), a theory test and a hazard perception test before they can take their chosen practical test.

Compulsory Basic Training

The first stage for anyone wishing to get a motorcycle driving licence is to complete the CBT. The Driving Standards Agency (DSA), in an attempt to reduce the crash rate amongst learner riders, introduced the CBT in December 1990 and it is now mandatory for all new riders wishing to ride a motorcycle, scooter or moped on public roads. The CBT consists of five Elements (DfT 2007b):

1. Introduction:
   - An eyesight test of reading a new style number plate at 20 metres or an old style number plate at 20.5 metres (DfT 2006b).
   - An explanation of what the CBT is and what will be taught.
   - A discussion of the correct use of safety equipment, including what is available, what protection it will give and how it should be maintained.
2. Practical on site training:
   - This section of the course is carried out in a private off-road area where basic information is given such as how to put the bike on its stand and how to start and stop the engine.
3. Practical on site riding:
   - This is also carried out in a private off-road area. Basic bike control skills are taught including: riding the bike in a straight line and stopping; changing gear; emergency stop; slow speed bike control; and performing U-turns.

4. Practical off-road training:
   - Skills that are needed for riding on the public road are taught in a private off-road area, such as hazard perception, observation skills and how to negotiate junctions.
5. Practical on road riding:
   - A minimum of two hours riding on public roads is required. During this period the instructor supervises the learner using a one-way radio.

If the course is completed to the satisfaction of the instructor then a DL196 certificate is issued validating the student's provisional motorcycle licence so that they can ride unsupervised on the road using a machine with power of less than 11kW (approximately 125cc engine size) that is equipped with 'L-plates', but they cannot carry passengers and they are not allowed to use motorways. The DL196 certificate is valid for two years. If both the theory and practical motorcycle tests have not been passed within this time, then for the rider to continue to use their PTW on the road, the CBT has to be retaken.

Theory and Hazard Perception Test

In order to take the practical riding test and obtain any class of full motorcycle licence ('P', 'A1' or 'A'), a theory and hazard perception test must be passed (DfT 2006d).

The first part of the theory test consists of 35 multiple choice questions about various subjects to do with riding, including road signs, maintaining a bike in a safe condition and safe riding techniques. Of the 50 questions asked, 43 must be answered correctly within the 57 minute time limit for the test to be passed. Since 14 November 2002 there has been a second part to the theory test, the hazard perception test. This comprises 14 video clips, each about 60 seconds in length, showing real road scenes where hazards develop. The rider is asked to identify the hazards; the faster the hazard is identified the higher the score obtained with a maximum score of five for each hazard. The pass score is 44 out of a maximum total of 75 points

Both parts of the theory test must be passed to obtain a theory pass certificate; this is valid for two years. The theory pass certificate and the CBT certificate (DL196) must both be presented at the practical test.

The Practical Riding Test

Once a DL196 certificate from passing the CBT and a theory pass certificate have been obtained then a rider can take a practical test on an appropriate machine for the licence type (DVLA 2003). For those attempting to obtain a 'P' licence the test must be sat on a moped. The 'A1' test must be taken on a bike of between 75 cc and 125 cc. The 'A' licence can be obtained through a normal route or via DAS. For the normal route the test must be taken on a machine between 120 cc and 125 cc, it must also be capable of attaining speeds greater than 100 km/h (60 mph); for a DAS test the bike must have a power output of at least 35 kW.

The practical test is conducted by a Driving Standards Agency (DSA) examiner who is in one-way radio contact with the rider throughout the test as he follows behind on his own bike. During the test some compulsory manoeuvres have to be carried out that includes a hill start, an emergency stop as well as pushing and riding the bike in a 'U-turn' (DSA 2004b).

A Brief History of the Riding Test and Learning to Ride

The manner of obtaining a UK riding licence that is in place today has evolved over nearly half a century and will continue to change with the implementations of EU driving licence directives. The first of these major changes occurred in 1960 when learners were restricted to bikes with a maximum capacity of 250 cc. Eleven years later in 1971 the minimum age for someone to ride any bike that was not classified as a moped was raised to 17, and this age limit is basically still in place today.

The next major change to the riding test was in 1982 when a two-part test was introduced. Prior to this, a rider obtained a full licence in a similar way to that of a car driver by passing one practical test. Part one was an off-road test and had to be completed before a provisional licence was awarded; this licence was only valid for two years. If the rider failed to complete the 'part two' on road practical part of the test within two years, then they would be banned from riding for a year. Further learner restrictions came into force in 1983 when learners were restricted to 125 cc capacity bikes.

The two-part test was amended in 1990 with part one being replaced by CBT. Further changes occurred in 1996 when a limit on bike size of 33 bhp was introduced for newly qualified riders in the first two years of their riding. The Direct Access Scheme (DAS), which could be used for older drivers to circumnavigate the two-year probationary period, was also introduced in this year.

The theory test, now a familiar part of the learning to drive experience within the UK, first surfaced in 1996 with the touch screen version making its debut in 2000. The hazard perception test was added to the theory test two years later in 2002.

The first EU Driving Licence Directive was implemented in 1996, however, there are now two further EU licence directives that are due for implementation.

EU Driving Licence Directives

The first EU Directive on driving licences introduced the principle that EU Member States should recognise each other's driving licences.

The Second Driving Licence Directive (2DLD) had some of its elements introduced in 1997, such as DAS. However, other key aspects of 2DLD concerned how the driving test was to be delivered and some of the manoeuvres to be carried out during the test. From October 2008 these aspects will be added: splitting the practical test into two parts with the first part taking place off-road and the second part being similar to the current practical test. To accommodate the new part one section of the driving test the Driving Standards Agency (DSA) has developed regional 'super test sites'. The off-road elements of the test include putting the bike on and off its stand; riding the bike in slalom and figure of eight patterns; carrying out a swerve at 50 km/h to simulate avoiding a hazard; a controlled stop; and undertaking a U-turn.

The Third Driving Licence Directive (3DLD) is due to pass into law in 2011 however, many motorcycling pressure groups have warned that this Directive will have a serious impact on motorcycling. Trevor Magner (2006) of the British Motorcycle Federation said:

> I fear the implications for the future of motorcycling are dire. Through its costs and complexity, this directive will be a big disincentive to anyone considering taking up motorcycling.

The Motorcycle Industry Association (MCIA 2007c) commented on the potential negative economic impact that this Directive could have:

> ... the economic impact of this Directive would have a seriously detrimental outcome in the UK for manufacturers, dealers, trainers and other sectors in the industry, with a predicted slashing of the motorcycle market by 50 percent over the decade following 2011.

So what are these changes that could destroy the future of motorcycling? There are four main changes: the minimum age for riding a machine with a capacity greater than 125 cc rises from 17 to 19; the Direct Access Scheme (DAS) will only be available to riders of 24 years or older – this option is currently available from 21 years of age; there will also be another licence category added ('A2') with a two-year period required between moving to a class that allows you to ride a larger machine; and a further practical test required to move between groups. One of the aims of this harmonisation is to increase motorcycle safety, a subject that is often closely related to the motorcycling activity.

Summary

This chapter has briefly explored some of the key issues, concepts and terminology that will be developed more fully in later chapters. Even in this brief introductory section, it is evident that while PTW riding may have some commonality with car driving, the complexities of the riding process, differences in manoeuvring and the vulnerability of riders make many road safety issues quite distinct from those of car drivers. Therefore the following chapter seeks to identify and highlight the specific road safety issues facing PTW users.

Chapter 2
Motorcycle Safety

Road Safety

The types of crashes that involve Powered Two-Wheelers (PTWs) differ from those experienced by other motorised road vehicles for various reasons. For example, the types of manoeuvres that motorcyclists can perform (for example, overtaking without crossing the centre line and filtering through traffic) are different as is visibility to other road users, and the performance of machines (for example, acceleration and cornering characteristics) is critically different. A study by Preusser et al. estimated that these factors contributed to 85 per cent of fatal PTW crashes (Preusser, Williams and Ulmer 1995). Mannering and Grodsky (1995) further discussed the differences of PTW crashes compared to other vehicles and gave a variety of reasons why the crash profiles differ. These were identified as:

- car drivers often only look for other cars as potential collision risks and therefore do not see bikes (looked but did not see);
- riding a PTW is a more complex task than driving a car;
- riding a bike may attract 'thrill seeking' individuals as it is considered more dangerous than other forms of transport.

These differences, along with the lack of protection afforded to motorcycle riders (RoSPA 2001), help to explain why PTWs are over represented in Killed and Serious Injury (KSI) crashes. This chapter examines the issues surrounding motorcycle safety including the nature and causes of crashes experienced by PTWs and measures to protect the rider.

PTW Crash Data

The *Association des Constructeurs Européens de Motocycles* (ACEM) commissioned a Europe wide in-depth study into the cause of crashes that involved PTWs – the Motorcycle Accident In Depth Study or MAIDS report (ACEM 2004). This comments that PTWs are different when compared to the majority of other forms of road transportation because bikes, along with their riders, are more sensitive to conditions. The riding of a bike is also a complex task that requires well-honed motor coordination and balance skills (Mannering and Grodsky 1995). Riding skills differ significantly from car driving skills, such as the use of independent front and rear brakes, weight distribution/shifting while riding and accelerating during cornering. Impairment by factors like fatigue or alcohol may therefore have a more significant effect on PTW riders than other vehicle drivers (Haworth and Rowden 2006), although there is no evidence to suggest that driving while impaired is a more prevalent problem among British riders.

With the control of a bike being more complex than that of a car and with PTWs being more sensitive to environmental conditions, it can be concluded that when things go slightly wrong this can quickly be amplified into a major incident. This is one reason why bikes may be considered more dangerous than cars. It is often stated that motorcycles have more crashes than cars, yet when FEMA reviewed the insurance statistics it showed that riders do not have a higher crash involvement risk than motorists (FEMA 2004), but as PTW users are more vulnerable, they have a higher risk of being injured or killed (Table 2.1). For car drivers the 1994–1998 average was lower than for PTW riders (11 per cent compared to 27 per cent). The percentage of car drivers involved in KSI crashes reduced to 8 per cent by 2005/2006, while for PTW users there has been little change.

The UK Government has set targets for reductions in all vehicle crashes. The aim is to reduce KSI crashes by 40 per cent by 2010 compared to the 1994 to 1998 baseline average (DfT 2000), with this target also applying to motorcycles. In the three-year review of the targets (DfT 2004c) it was reported that good progress was being made towards this target, except in the case of PTWs where there was an increase of 16 per cent in KSI crashes. This increase was put down to exposure, as when PTW crashes were related to distance travelled, there was a reduction in the crash rate.

Table 2.1 KSI/slight crashes

		1994–1998 average		Oct 05 to Sep 06	
		n	%	n	%
Car	KSI	23,254	11%	14,480	8%
	Slightly injured	180,034	89%	159,870	92%
	All casualties	203,288	100%	174,350	100%
PTW	KSI	6,475	27%	6,370	27%
	Slightly injured	17,547	73%	17,080	73%
	All casualties	24,023	100%	23,450	100%

Source: Transport Tends 2006, DfT 2007.

PTW Crash Causes

With the exception of pedestrians, when motorcyclists are involved in crashes they are more likely to suffer serious injuries than other road users. Their injuries are more likely to be causing problems a year after the crash than injuries suffered by other road users; again with the exception of pedestrians (Mayou and Bryant 2003). Therefore the issue of motorcycle safety is one that is taken seriously within the motorcycle community.

Sexton, Fletcher and Hamilton (2004) surveyed motorcyclists to look at the relationship between crash risk and other variables. This showed that those who rode smaller bikes of less than 125 cc were 15 per cent more likely to have crashes than those riding the larger machines, although the larger machines were more likely to be involved in fatal crashes. This research confirmed that the risk per mile of a fatal crash increases with engine size (Sexton, Baughan, Elliot and Maycock 2004). Not surprisingly the report also showed that the crash risk increased with the number of miles ridden, that is, with exposure. Rutter and Quine (1996) also found that, after taking into account exposure rates, younger motorcyclists are more likely to be killed or injured on the roads. A similar finding was reported by Yannis, Golias and Papadimitriou (2005), who also stated that although rider age was a factor in PTW crashes, the engine size of the machine being ridden was not significant, a finding that concurred with Langley, Mullin, Jackson and Norton's (2000) research. It may be that engine size is not related to the crash rate but may relate to KSI crashes as these bikes have the capability to travel further, and faster, than smaller bikes (Sexton, Fletcher and Hamilton 2004).

Speed will always be an issue as the resultant energy (E_k) of an impact is related to the mass of the object (M) and velocity (V) squared. In mathematical terms the relationship is expressed as $E_k = ½(MV^2)$ (Aarts and Van Shagen 2006). As the equation indicates, speed (velocity) substantially increases the amount of energy created which means the impact of the crash will be much harder. Therefore for motorcycles involved in crashes, the risk and severity of injury increases with speed. Most motorcycle crashes though happen at slow speeds (RoSPA 2001); in over 70 per cent of cases, the PTW impact speed is less than 30 mph (ACEM 2004). However, the statistics show that a majority of KSI crashes occur in non-urban areas. Again this is most likely to be related to the fact that these are areas where higher speeds can be obtained. In the past it was often suggested that these 'high speed non-urban crashes' were a bike problem, however, 'The Key 2005 Road Accident Statistics' (Scottish Executive 2006) shows that within Scotland a greater percentage of car drivers have crashes in non-urban areas compared to motorcycle riders (Table 2.2 and Table 2.3); Table 2.4 shows the percentage of non-built-up crashes out of all crashes (built-up + non-built-up). As speed is perceived as the reason why there are more KSI crashes on non-built-up roads it is interesting to note the higher percentage of KSI for cars over bikes on this type of road.

Table 2.2 Comparison of crashes for cars and PTWs occurring on Scottish built-up roads

	PTW					
Year	2003		2004		2005	
Killed	12	2%	5	1%	3	<1%
KSI	159	27%	146	28%	151	29%
All	591	100%	527	100%	572	100%
	Car					
Year	2003		2004		2005	
Killed	22	<1%	28	<1%	20	<1%
KSI	497	9%	376	74%	342	71%
All	5381	100%	5153	100%	4828	100%

Source: Scottish Executive 2006.

Table 2.3 Comparison of crashes for cars and PTWs occurring on Scottish non-built-up roads

	PTW					
Year	2003		2004		2005	
Killed	38	7%	36	8%	31	6%
KSI	258	49%	244	53%	244	48%
All	523	100%	461	100%	506	100%
	Car					
Year	2003		2004		2005	
Killed	162	3%	139	2%	133	2%
KSI	1,194	19%	1,199	19%	1,082	18%
All	6,359	100%	6,418	100%	6,102	100%

Source: Scottish Executive 2006.

Table 2.4 Percentage of crashes for car and PTW occurring on Scottish non-built-up roads

	PTW			Car		
Year	2003	2004	2005	2003	2004	2005
Killed	76%	88%	91%	88%	83%	87%
KSI	62%	63%	62%	71%	76%	76%
All	47%	47%	47%	54%	55%	56%

Source: Scottish Executive 2006.

When comparing non-urban areas to urban areas the number of PTWs having collisions with cars decrease from 64.1 per cent to 46.7 per cent; there is a small increase in collisions between PTWs (6.3 per cent to 9.6 per cent) and also a substantial increase from 4.2 per cent to 19.7 per cent for crashes between PTWs and fixed objects (ACEM 2004). With a higher KSI rate in non-urban areas as well as a different crash profile, there is an argument for treating urban and non-urban crashes separately for research and safety intervention purposes.

The MAIDS report (2004), which examined 921 crashes involving PTWs, found that in 50 per cent of crashes the primary contributing factor was human error on the part of the other driver with 70 per cent of these errors being failure to perceive the bike – a 'looked but did not see' error (ACEM 2004). In similar research, Mannering and Grodsky (1995) found that 'drivers not being attentive' was a main cause of the crash rate for motorcycles. The MAIDS (ACEM 2004) report found that in the majority of PTW crashes the bike collided with another vehicle (80.2 per cent) and that a passenger car was the most frequently collided with object (60 per cent). Over half of all PTW crashes occur at junctions. These figures suggest that the causation of crashes is complex but identifies that bikes not being seen by other road users is a major problem.

Age and experience also have an effect on crash rates. The MAIDS report states that there is a lower risk of being involved in a crash for riders in the 41–55 age group (ACEM 2004), with the 18–25 age group being over represented (Chesham, Rutter and Quine 1993). It is often stated that the 'born again' bikers, who mainly fall into the 41–55 age group, are the main PTW crash problem. While it is true that in absolute numbers the 41 to 45 age group do account for a large proportion of those having crashes, it is also true that they form the majority of those who ride. When this is taken into account this group actually has a lower risk. The 40 to 49 year olds have 18 per cent of the total KSI crashes, however, this group makes up 25 per cent of the riding population. This is illustrated in Table 2.5, the rider percentage is a 2002 to 2004 average, with KSI figures from 2004 (DfT 2007f).

Riders with less than six months' experience are more likely to be involved in a crash when compared to the rest of the riding population. These riders are more

Table 2.5 Age of riders and KSI

Age	Percentage of riders	Percentage of KSI
16–19	10%	19%
20–29	10%	23%
30–39	27%	27%
40–49	25%	18%
50–59	17%	8%
60+	10%	4%

Source: Compendium of Motorcycling Statistics 2006.

likely to make decisions or manoeuvres that result in a crash, suggesting that rider experience is useful for developing skills in risk identification and anticipation of dangerous situations (ACEM 2004). Inexperience can be linked to youth with those who are just starting out on their life as a powered road user being more likely to be involved in crashes than other drivers and riders (Stradling 2005). The other less experienced groups consist of those who are either returning to riding after a long lay-off or taking up biking in later in life. Despite this group being more likely to have driving experience, and the road craft skills that comes with that, they will not have any recent riding experience, and as riding is significantly different to driving (Mannering and Grodsky 1995) this experience will only be partially useful.

Although there is no substitute for experience, training can help to bridge the gap between a novice and experienced rider. As noted earlier, the riding of a motorcycle is more complex than that of a car (Mannering and Grodsky 1995), particularly for skills specific to motorcycle riding such as using independent front and rear brakes. The Royal Society for the Prevention of Accidents (RoSPA) (2001) reported that the correct use of brakes could prevent 30 per cent of crashes, showing an area where more training would be beneficial, hence training, or other interventions, could be useful in reducing KSI crashes amongst PTW users.

Crashes and Riding for Pleasure

There are differences between the types of crashes that motorcycles have when compared to other road vehicles. This is illustrated by using official crash data, known as STATS19 data; this dataset was examined for Scotland using 2005 data.

Motorcyclists are far more likely to have a crash while negotiating a bend or overtaking (Table 2.6). Many riders when riding for pleasure report that bends are a significant enjoyment generator (Broughton 2008a). Therefore riders could seek out bends more than other road users. However, Broughton (2007) also suggests that task difficulty influences this type of crash. Overtaking crashes are also related to task difficulty with the nature of bikes exacerbating this, as the manoeuvrability of PTWs means that motorcycles can overtake in places where cars and other vehicles cannot. Issues of task difficulty and its impact on riding are explored in more depth in Chapter 9.

The seeking of bends for riding pleasure is also reflected in where crashes occur, with motorcycles being more likely to be involved in a crash in a rural setting than other vehicles (Table 2.7).

Another element of riding for pleasure is where crashes are happening with reference to the rider's domicile with riders significantly more likely to be crash involved at greater distance from home. Table 2.8 shows that nearly 20 per cent of motorcycle crashes occur more than 40 km from the rider's home, compared to only 12 per cent of other road users. This is almost certainly a reflection of the different patterns of travel for PTW users (DfT 2007c).

Table 2.6 Comparison of crash manoeuvres for PTWs and other vehicles

	PTW	Other	All vehicles
Ahead Bend	21%	12%	13%
Ahead Other	52%	49%	49%
Overtaking	11%	3%	4%
Other	16%	36%	35%
Total	100%	100%	100%

Source: STATS19 for Scotland 2005.

Table 2.7 Comparison of urban and rural crashes

	PTW	Other	Total
Urban	55%	62%	62%
Rural	42%	32%	32%
Motorway	3%	6%	6%
Total	100%	100%	100%

Source: STATS19 for Scotland 2005.

Table 2.8 Crash distances from domicile

	PTW	Other	Total
Under 40 km	81%	88%	87%
Over 40 km	19%	12%	13%
Total	100%	100%	100%

Source: STATS19 for Scotland 2005.

Given that the nature, type, location and severity of PTW crashes are quite different to other road users such as car drivers, then interventions may be more effective if they recognise this.

Interventions

A review of fatal motorcycle injuries in South East Scotland (Wyatt, O'Donnell, Beard and Busuttil 1999) found that injuries to the head, neck and chest were the most severe. The nature of these injuries led them to conclude that crash prevention and injury reduction measures are the best methods for reducing rider deaths, rather than improvements in the treatment of injuries. This section discusses some of the interventions that aim to reduce the number of crashes.

Training

When examining the reasons for crashes, an argument can be made for a high level of training for PTW users. In a third of crashes examined in the MAIDS report (ACEM 2004), the PTW rider had adopted some faulty traffic strategy that contributed to the crash, such as approaching a corner too fast. This suggests that additional training could be provided in the selection of correct traffic strategy.

Compulsory Training

Currently anyone who is taking up biking in the UK must take a 'Compulsory Basic Training' (CBT), a short training course at an approved school. This course consists of a mix of theory, off-road practice and some time to practise the newly learnt skills on the public road. Once the CBT has been passed then a bike of up to 125 cc can be ridden on the road with learner restrictions (DfT 2004b). Further training is normally taken to enable the rider to pass the required tests. They are then able to use a bike without the learner restrictions, although some engine size restrictions may still be imposed depending on the rider's age and the type of bike used to take the test (DfT 2005a). A fuller description of the procedure to obtain a PTW licence in the UK was given in Chapter 1.

Compulsory Basic Training is undertaken for the purpose of obtaining a licence and the majority of it is carried out in an urban environment. While this is where most crashes happen, it is not the place where most KSI crashes happen (DfT 2007c). The MAIDS report found that in a fifth of PTW crashes there was only one bike involved and that the rider was travelling at speeds over 60 mph (100 km/h) on non-urban roads. Currently some motorcycle trainers may not develop these skills in their pupils as there is only limited, if any, testing of these skills during the practical riding test. There has been a trend towards more fatalities involving higher travelling speeds. In general the impact speeds for single vehicle crashes are higher than for crashes that involve other vehicles (ACEM 2004). Lack of control can also be a problem, with 'running wide on a turn' being the most common type of loss of control (23.0 per cent); 'braking slide-outs' on the low side (14.5 per cent) and 'low side cornering slide-outs' (11.0 per cent) being further main factors (ACEM 2004). Additional training, and training on non-urban roads could help to reduce these kinds of crashes and reduce the KSI figures. Note that the 'low side' of the bike is the side that is leaned towards the ground while cornering, for example, in a right turn, the right side is the 'low' side and the left side the 'high' side, because it is higher off the road surface. A 'low-side' crash is when the tyres slide from under the bike and the bike lands on its low side.

Other Training

There is currently a variety of voluntary training schemes in addition to those needed to obtain a licence. The Motor Cycle Industry Association (MCIA) (2006b) reported on a survey showing motorcycle riders are positive about training and that the most popular training organisation was the Institute of Advanced Motorists (IAM). The Bikesafe scheme was also popular. Bikesafe is a scheme that is run by police forces around the United Kingdom, using police motorcyclists to pass on their skills and experience (Motorcycle UK Ltd 2007). Research into the effectiveness of this scheme showed that riders believed that it was useful and that they would recommend it to a friend. One of the conclusions of the report was that there 'may be a need for a greater focus on attitudes to riding as part of the Assessed Ride programme' (Ormston, Dudleston, Pearson and Stradling 2003). Advanced training can take many forms, often leading to a recognised qualification, such as that issued by IAM or RoSPA. Many insurance companies will offer reduced premiums to riders who have undergone this type of training. According to RoSPA (2007) advanced riders are 20 per cent less likely to be involved in a crash than those who are not so qualified. In 2000, 118,853 riders underwent training, compared to 90,656 (60,008 passed) taking their practical riding test (DfT 2006a) – this figure compares to a total number of motorcyclists of around 1,215,000 (DfT 2007c), therefore around 10 per cent of riders take voluntary additional training. The MCIA (2006b) report stated that of the riders that haven't participated in any training, 37 per cent said that lack of time was the reason, and 21 per cent indicated that cost was a barrier. This indicates that the majority of riders have a willingness to participate in the right circumstances.

Training does not have to be formalised. Opportunities can be taken to modify rider behaviour while safety is in the forefront of the rider's mind, for example, after a crash while receiving treatment for their injuries by medical staff. A nurse's negative attitude while treating a motorcyclist does not create an atmosphere that is conducive to educating the rider, but if the nursing staff understand the problems facing riders then they can, at the correct time, use evidence-based statements in an attempt to modify the rider's future behaviour (Blanchard and Tabloski 2006).

Training is an area that is being used to try and reduce the number of rider casualties, but there is a need to underpin skills training with the reasons why riding has to be done in a certain manner and the consequences when it is not. Skills training alone can actually increase the risk of the rider being involved in a crash due to an over-estimation of skills (Rutter and Quine 1996). The frequency of training should also be considered as motorcycle training may only have short-term effects. Goldenbeld, Twisk and de Craen (2004) found that the effects of PTW training could not be differentiated from a group with no training after a period of 11 months.

While extra training for PTW users would be beneficial, especially in the skills needed for non-urban riding, this would only be addressing part of the problem as other road users also create a risk to riders. For instance, in 40.6 per cent of crashes the other (non-bike) vehicle had adopted some faulty traffic strategy that contributed towards the crash, with over 70 per cent of 'other driver errors' being failure to see the PTW. Other vehicle drivers who hold a PTW licence are more likely to see a PTW, which shows that with some training this type of crash could be reduced (Clarke, Ward, Bartle and Truman 2004; Crundall, Bibby, Clarke, Ward and Bartle 2008; Mannering and Grodsky 1995). Car drivers need to be educated so that they are made more aware of the needs of PTW users, and their vulnerability (RoSPA 2001). In the 1970s the UK Government launched a public information film for this purpose, with the slogan:

> 'Think once, Think twice, Think bike'. (Central Office of Information for Department of Transport 1978)

RoSPA (2001) commented that 'The slogan "Think Bike" is as relevant today as it ever was.' The 'Think Bike' message has been updated, with the current version entitled 'Think – take longer to look for bikes' (DfT 2006e). Even if a higher bike awareness is achieved, it is unlikely to completely eradicate the problem of other road users not seeing bikes, therefore it is up to riders to become more defensive in their riding styles and to afford themselves as much protection as possible while riding.

Protecting the Rider

Although the nature of PTWs and the vulnerability of their users do not allow riders to protect themselves as comprehensively as car drivers, there are a number of steps that can be taken to reduce injury in the event of a crash and improve the likelihood of surviving a serious incident. The use of protective clothing and, in particular, helmets can reduce the possibility of serious head and brain trauma.

Motorcycle helmets have been proven to be effective in injury reduction for riders involved in crashes (American College of Surgeons 2004; Branas and Knudson 2001; Kraus, Peek, McArthur and Williams 1994; McGwin, Jr., Whatley, Metzger, Valent, Barbone and Rue 2004). There is an urban myth that helmets can make riding more dangerous as they affect the ability of a motorcyclist to see and hear. However, research by McKnight and McKnight (1995) showed that the reduction to vision and hearing is small and only has a minimal negative effect on safety. The protective benefits afforded to the rider compensate for this minimal addition to risk.

Within the UK it has been compulsory for riders to wear a helmet since 1973; the only exception was brought in in 1976 to allowed exemptions for followers of the Sikh religion when wearing turbans.

Nowhere is the battle over the compulsory wearing of helmets better illustrated than in the United States of America where until 1997 the wearing of helmets was mandatory in most states as the federal government withheld highway money from states without mandatory helmet laws. However, since congress rescinded this policy, various states have repealed or weakened their helmet law due to pressure from free-choice groups (*USA Today* 2008). Currently 20 states and the District of Columbia require a helmet to be worn. Laws that only require some, such as those under 18, to wear a helmet are in force within 27 states. Three states, Illinois, Iowa and New Hampshire have no helmet use laws at all (Insurance Institute for Highway Safety 2008). This situation is very fluid with many states considering their position on helmet laws.

The changes in the helmet law allows the effectiveness of helmets to be examined. Vaca (2006) compared the increase in fatal motorcycle crashes before law repeal in 1997 to post repeal in 2003. The data showed that there were increases in all five states examined, ranging from a 101 per cent increase in Florida to a 290 per cent increase in Louisiana. An article in USA Today discussing helmet laws stated that:

> When mandates are repealed, deaths and injuries increase. (*USA Today* 2008)

In states where helmet use is not compulsory, the take up is limited. In a meta analysis conducted by Elvik (2004) it was found that compulsory helmet use reduced the number of injuries to riders by between 20 per cent and 30 per cent with the same study showing that repealing of compulsory helmet laws resulted in 30 per cent more deaths. Opposers of helmet laws dispute figures and suggest that the rise in fatalities is purely due to increasing numbers of registered motorcyclists; they believe that rider education and driver awareness raising are better ways to reduce crashes (Hennie 2008).

In the spring of 2008 the Department for Transport (DfT) within the UK introduced a rating system for helmets called SHARP (Safety Helmet and Assessment Rating Programme) with helmets being rated for safety from one to five stars. Even though all helmets have to comply with the current standards there is a variation in the exact amount of protection that a helmet can give. This rating system gives potential users an independent assessment of the safety performance of each helmet so that they can make an informed decision when purchasing a helmet. Research from the DfT suggests that up to 50 riders lives could be saved each year within the UK if helmets of a high rating were used (DfT 2007e).

Helmets come in two basic designs: open face and full face, with two types of full face helmets: normal and flip (Figure 2.1). A full face helmet has a bar around the chin and when the visor is closed the whole head is covered. A flip face helmet is very similar to the full face helmet, however, the front of the helmet can 'flip-up' to expose the face but the front should be in the down position when the user is riding. The flip face helmet allows the user to expose their face when they are not riding, this can be useful for activities such as being able to purchase fuel without fully removing the helmet. An open face helmet does not have a bar across the chin therefore the face is not fully enclosed. Crash investigations show that full face helmets give more protection than the open face variety (Mills 1996).

The visor of a helmet is critical as any scratches or blemishes can impede the rider's vision. It is permissible for a helmet to have a certain amount of tint, but

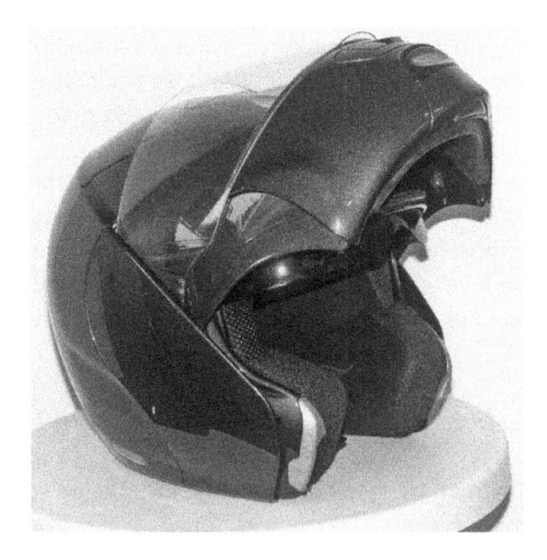

Figure 2.1 Full-face helmet, flip design

they must comply with BS4110:1979 (as amended) or UNECE Regulation 22.05. These standards state that visors that are intended for use in the daylight must allow a minimum of 50 per cent light transmittance. However, a trawl through online rider forums will show that many riders wear visors significantly darker than the standard and also argue that the standards are flawed (DfT 2002).

Other Safety Equipment

Within the UK an approved helmet is the only compulsory piece of safety equipment that has to be worn; however, other protective equipment can also help in reducing the severity of rider injuries (Elliott, Baughan, Broughton, Chinn B, Grayson, Knowles, Smith and Simpson 2003). All motorcycle safety equipment must comply to EN 1621 (British Standards 2003). If it does, it is entitled to carry a CE mark to demonstrate this. The EN 13595 standard for motorcycle equipment only applies to safety gear being used by professional riders (British Standards 2002). Ensuring that the protective clothing complies with the correct standards is essential as some manufacturers make fashion clothing that looks similar to protective clothing.

All properly designed safety equipment can, and does, reduce injuries to riders however, there is concern that some Government and European Union policies discourage use as VAT is chargeable on all motorcycle protective equipment with the exception of the helmet. There is even the anomaly that a helmet bought with a visor is VAT free but a replacement visor attracts the higher rate of VAT (HM Revenue and Customs 2002).

The use of headlights during the daytime has been adopted by most motorcycle riders to improve their visibility, however, the effectiveness of daytime running lights for motorcyclist is unproven. Elvik, Christensen and Olsen (2003) reported a reduction of 32 per cent for multi-party daytime crashes, but the range of the results making their conclusions inconclusive (95 per cent CI: -64 per cent – 28 per cent).

There are other safety issues not directly associated with crashes, for example, noise induced hearing loss. Although to date there are few studies on hearing loss associated with recreational PTW use, some work has been undertaken on occupational motorcyclists such as police, professional racers, paramedics, couriers and pizza delivering riders. A study by Jordon, Hetherington, Woodside and Harvey (2004) found that the noise exposure caused by high levels of aeroacoustic noise posed a risk of developing noise induced hearing loss:

> Noise levels in excess of 105 db(A) were recorded for motorcycles travelling at 70 mph, the maximum public roads speed limit in the UK. (Jordon, Hetherington, Woodside, and Harvey 2004: p. 66)

While this study related to occupations which may require regular exposure to high levels of noise, the issues of noise are also relevant to other PTW users. The use of conventional earplugs was found to have limited use due to the high levels of low frequency noise that dominate the wind noise spectrum. Neck seals, although inconvenient and difficult to fit properly, did offer a level of protection as the dominant noise source came from the space between the rider's neck and the chin bar.

Safety Equipment on PTWs

Other safety devices have also been incorporated onto bikes, however, due to the nature of the mechanical design of PTWs and rider ergonomics, there are fewer available for PTWs than are available for other vehicles. For example, PTWs cannot have side impact bars or seatbelts. Another difference with PTWs is their lack of stability when compared to a car, for this reason ABS braking systems that work on the principles of a car would be unsafe on a bike.

On modern bikes, with their relatively high performance, an aggressive front brake is required. This, however, has the potential effect of the rider easily locking the front wheel during heavy braking. This causes the bike to become unstable and gives rise to a high risk of the rider being involved in a crash. Advanced, or anti-locking, brakes can prevent the locking of the front wheel and thus aid in reducing crashes by an estimated 10 per cent (Sporner and Kramlich 2000). However, due to the complex nature of motorcycle ABS, they tend only to be fitted to the more expensive machines. There is also resistance from some riders to using a bike fitted with ABS as this is seen as removing some of the control from the rider.

Air bags have now started to appear on motorbikes, with Honda introducing one on the GL1800 Goldwing in 2006 (see Figure 2.2). Airbags for PTWs are more difficult to implement than those designed for cars, as when a PTW is involved in a crash, the nature of the interaction of rider and bike is more complex than the interaction between a car and a driver. For one thing, a car driver is often strapped into the car and is not liable to be projected from the vehicle. When a PTW is involved in a head-on crash the rider continues to be propelled forward and collides with the collided object at a speed approaching the original impact speed. Happian-Smith and Chinn (1990) demonstrated that at speeds above 30 mph it was not possible to prevent the rider being ejected, however, reduction of ejection speed could still have safety benefits.

Riders suffer leg injuries in about 80 per cent of all crashes (Huang and Preston 2004) mainly because the legs are exposed on either side of the bike. There has been a lot of research effort put into finding ways to protect a rider's legs. Mechanical and 'air bag' leg protector devices have been developed (Chinn and Hopes 1985); it has been estimated that leg protectors can reduce leg injuries suffered in a bike crash by 50 per cent (Sporner Langwieder and Polauke 1990). The moves in the early 1990s to make leg protectors mandatory on new bikes

Figure 2.2 Honda airbag system

was not successful, partly due to campaigning of biking pressure groups. These pressure groups argued that there was an additional risk posed by leg protectors in a crash as legs could become trapped between them and the road rather than the rider sliding off. There was also a problem in ensuring that the protectors could accommodate riders of differing shapes and sizes

Other safety devices, such as traction control systems that prevent the back wheel spinning, and hazard warning lights are now available on some bikes. The very nature of motorcycles means that safety features that are standard within other vehicle types cannot be directly transferred; however, technology is gradually pushing its way into bike safety and more safety innovations are becoming available.

Road Design and Layout

There are considerable differences between how motorcycles interact with the road and road environment compared to other vehicle types. If the design of roads does not take this into account then motorcyclists' lives can be put at risk, for example:

- Road furniture that is not positioned to take into account the overhang of PTWs (ETSC 1998; RoSPA 2001; VicRoads 2001).
- Road design and maintenance being aimed at non-two wheeled vehicles (Institute of Highway Incorporated Engineers 2005) – for example, metal covers and road paint that give no traction in the wet (see Figure 2.3).
- Potholes and longitudinal roadway ridges, mainly caused by HGVs. Road defects are a contributing factor in 3.6 per cent of crashes (ACEM 2004).
- Roadway debris (FEMA 2004).
- Diesel spillages (BMF 2004; BMF 2005).
- Traffic calming measures that are not suitable for PTWs (RoSPA 2001).

Road design to minimise motorcycle risk is important if the number of fatalities are to be reduced. The Institute of Highway Incorporated Engineers (2005) has

Figure 2.3 Example of metal on the road surface

issued guidelines for road design, maintenance and policy to improve PTW safety. PTW riders are more vulnerable than car drivers and have more complex tasks to undertake in order to propel their vehicle, so specific measures may be needed to reduce KSI numbers. Measures aimed at the majority of road users, such as car drivers, may not always be sufficient for PTW users.

Policy and Legal Interventions

Interventions by government are often linked to other types of intervention. For example, governments have changed the methods for getting a 'full' riding licence over the years to compel potential riders to take training before getting onto the open road. As explained earlier (see Chapter 1), in the late 1970s it was permissible for anyone aged 17 or over who was in possession of a provisional driving licence to get onto a bike and ride it on the open road. This riding could be undertaken without any training and the rider may have had no idea about the complexities or mechanics of riding, let alone road craft skills. The introduction of CBT ensured that all riders at least had some training prior to being admitted to the public highway.

Legal changes have also changed the types of bikes that can be ridden. In the late 1970s and early 1980s a learner could ride a motorcycle of up to 250 cc. Around 1980 Yamaha launched the RD250 LC motorcycle; many people believe it was this bike that forced the Government to act to reduce the engine size and power of bikes that learners could use as the RD250 was capable of doing over 100 mph and produced 35 bhp.

The implementation of the second and third European Union (EU) Driving Licence Directives has also had an effect on the overall KSI numbers for PTW riders. Part of this is due to the safety measures incorporated into the legislation, such as the reduction in power of the bike that a learner can use and the increase in training required to ride. Some of the reduction in KSIs may be due to a reduction in the number of riders due to the increased level of training required and a greater restriction on the type of bike they can ride. The harder it is perceived to be to obtain a licence to ride the type of PTW that they wish to ride, then the lower the likelihood that people will take up riding. For example, the two-part test that is planned to be implemented in 2009 will make it more expensive, and difficult, to obtain a licence and it is forecast that the number of new riders will decrease.

There are concerns that the new test criteria planned to be implemented in 2009 will impact on the numbers of licenced riders. Not only will it increase the cost of gaining a licence, the limited number of centres will make it hard for those in rural areas, or island communities, to reach test centres. While it would appear to the consensus that better training and the addition of more challenging manoeuvres is a positive thing, there is concern at the way it is being introduced. In an article for the Highland Correspondent, Ross (2008) raised concerns that the inability of many to reach centres may lead to some riding without a licence. Alan Reid, The

Liberal Democrat Member of Parliament (MP) for Argyll and Bute was quoted as saying:

> Better testing may save lives, but only if testing is accessible to learners. (Ross 2008)

As well as ensuring that training is taken before a licence is obtained, in some situations riders can be required to undertake courses by the courts. These schemes operate on similar lines to the Driver Improvement Scheme (Association of Chief Police Officers 2003) with the training being offered as an alternative to prosecution for Section three offences. Section three offences include careless driving and driving without reasonable consideration (Crown Prosecution Service 2006). The aim of the course is to change the attitude and behaviour of these errant riders (DfT 2005a) and not to increase riding skills. This would suggest that a real benefit to motorcycle safety would be to adjust the rider, and to do this some understanding of the psychology of the rider has to be understood. Sudlow (2003), in a report written for the DfT on motorcycle training schemes, concluded that to train a rider properly it is important to understand the rider and the motivation of riding.

Summary

Given the relative risk posed by PTWs as a mode of transport, safety is a major concern. Measures such as wearing helmets and appropriate clothing can offer some protection but the nature of PTWs makes further protection measures more complex. Compulsory training and restrictions on riders form part of the solution but there is a need for ongoing training for riders that goes beyond the teaching of skills and also deals with riding attitudes and motivations. An increased awareness amongst other road users is also needed to reduce crash rates.

In order to allow any interventions the greatest possible chance of being effective, the circumstances, motivations and behaviours surrounding PTW use must be understood. The following chapter explores some of the issues concerning risk by examining a number of theories and their relevance for PTW users.

Chapter 3
Task Homeostasis Theory

The riding of Powered Two-Wheelers (PTWs) gives rise to a higher level of risk when compared to other forms of transport. A rider's likelihood of being involved in a Killed or Seriously Injured (KSI) crash is much higher than for a driver with some of this additional risk being due to riding being more complex than driving (Mannering and Grodsky1995). In the UK and elsewhere there have been many interventions to make riding safer; for example, legislation to make the wearing of helmets compulsory. However, to understand the effect that these interventions have, it is important to understand how risk, and other relevant factors, relates to rider behaviour. There are a range of risk theories but this chapter will focus on the two theories most prevalent in the field of road safety; Risk Homeostasis Theory (Wilde 1982), which is often referred to as Risk Compensation Theory (although there are some differences in the details between these two formulations); and Task Homeostasis Theory (Fuller 2005).

Risk Homeostasis Theory

The theory of risk homeostasis posits that an individual has an inbuilt target level of acceptable risk and this does not change over time but does vary between individuals. When there is a change to the risk levels that an individual is faced with, then that individual will change their behaviour so that the level of risk is restored back to their target level.

One of Wilde's arguments is that anti-smoking campaigns and safety interventions within the work place are ineffective. He also suggests that road safety interventions such as air bags and compulsory helmet use have had little or no effect on the overall crash rate and the costs of these crashes.

The basic thread of the Risk Homeostasis Theory is that the overall crash and injury rate remains comparatively constant despite, and regardless of, safety interventions. For this to happen within road safety all road users have to collectively change their behaviour to keep the risk level, and crash rate, constant. Wilde explains that this happens in a similar way to a thermostat with behaviour changing to be more, or less, risky depending upon where the risk level is set on the thermostat. One of the commonly quoted 'proofs' of Risk Homeostasis working is:

> I once heard said in support of risk homeostasis theory: the safest possible cars would have no seatbelts and a large spike affixed to the centre of the steering

> wheel. Although they might be safer, such a design would no doubt add considerably to journey times. (Adams 1995; *British Medical Journal* 2002)

The examination of this theory is appropriate within this book, as it has been quoted by motorcycle lobby organisations to support their point of view on certain policies. For example:

> The advantages of wire rope barrier systems have been advocated through concepts such as risk compensation. (Motorcycle Action Group 2005)

> Although risk compensation may negate some safety benefits, we consider that the manufacturers should make Anti-lock Braking Systems (ABS) available as an option for more of their model ranges including the smaller machines. There should be no question of considering making ABS mandatory at this stage. (BMF 2007)

It is also relevant to discuss it as the proposer of Risk Homeostasis said that:

> The theory of risk homeostasis (also known as 'risk compensation') was primarily developed and validated in the area of road safety. (Wilde 1998)

However, despite Risk Homeostasis Theory often being discussed with reference to road safety issues, it has also attracted significant criticism and is not widely accepted within the road safety expert fraternity. For example, O'Neill and Williams (1998) stated that:

> These so-called theories that purport to explain human behavior in the face of risk are nothing more than hypotheses with a large body of empirical evidence refuting the studies that allegedly validate them. (O'Neill and Williams 1998)

According to Risk Homeostasis Theory if, for example, a rider started to wear a crash helmet when riding that this will have the affect of protecting the rider's head and therefore reducing the risk to life; therefore the motorcyclists will ride in a harder and more reckless manner because of the lower risk of death or serious injury. This new 'less safe' activity will be without cognisance of any damage to the bike that may be caused by a crash as the only issue of importance will be the risk of injury. This is unlikely as risk takes into account all consequences, not just physical harm but also loss of financial and other resources.

The theory of 'Risk Homeostasis' has not changed over the quarter of a century since its inception. Some of the theory's supporters argue that this is because the data fits the theory (Evans 1986). Wilde in support of this position states that:

> ... airbag equipped cars tend to be driven more aggressively and that aggressiveness appears to offset the effect of the airbag for the driver. (Petterson, Hoffer and Millner 1995)

There have been many studies into the effects of airbags; the data does not support Wilde's position as they show that airbags have actually reduced the number of car crash-related deaths (Kahane 1996; O'Neill and Lund 1992).

There may be several reasons for this absence of change to the theory of Risk Homeostasis; one might be that it represents a theoretically coherent picture that adequately fits the data. However, there have been a number of challenges to this proposed theoretical coherence (Evans 1986; McKenna 1985; McKenna 1988).

The idea that the population acts as a collective to keep the crash and injury rate at a constant level is part of the hypothesis with Risk Homeostasis Theory. For this auto-adjustment system to operate there must be some feedback, similar to the feedback that a thermostat uses to control temperature correctly, to the collective so that they can change their behaviour in a manner that is consistent with the theory. This raises the awkward question of how does this feedback occur? Wilde explains that:

> Each action carries a certain level of injury likelihood such that the sum total of all actions taken by people over one year explains the accident rate for that year. This rate, in turn, has an effect on the level of risk that people perceive and thus upon their subsequent decisions, and so forth. (O'Neill and Williams 1998)

For this proposition to work all road users must take notice of the current crash rate and adjust their behaviour so that the collective rate is manoeuvred back to the target rate. Wilde uses the measurement of crashes, and severity of crashes, per road user capita, however, this measurement does not take into account the increased exposure due to the increase in distances being driven.

If one was to accept Risk Homeostasis Theory as being valid then one would also have to accept that the only way to improve road safety is to instil the desire to be safe into road users. This is not to say that instilling the desire to be safer will not have positive effects on road crash statistics, it almost certainly will, but the evidence shows that other methods are also effective. If it is acknowledged that the only way to improve road safety is to instil the desire to behave in a safer manner within all road users then safety measures such as the compulsory use of motorcycle crash helmets would not have the desired effect of increasing rider safety as riders would only ride harder to compensate for the increased protection gained from the helmet.

The United States of America provides an opportunity to examine data in a 'before and after' experiment. As discussed in Chapter 2, nearly all the states had laws mandating helmet use around 1965, however, many states in the following years have repealed or altered these laws making helmets no longer a legal requirement. After the repeal of compulsory helmet law within a state the number of riders taking advantage of the protection of a helmet also dropped and with this came an increase in rider deaths in the order of 15 per cent to 40 per cent suggesting that riders do not ride in a safer manner when they are not wearing a helmet (Evans 1991; Evans 1994; Flemming and Becker 1992; Muelleman, Mlinek and Collicott 1992; Watson, Zador and Wilks 1981).

The evidence given here questions the validity of Risk Homeostasis Theory, with many studies showing that the data on road traffic crashes does not support the theory. There has never been any systematic review of data that could show that Wilde's theory is valid (Thompson, Thompson and Rivara 2001). Evans (1991) expresses the view that:

> ... the tone of advocacy for the claim has been largely philosophical, metaphysical, and theological in nature, unencumbered by the standards, methods, or norms of science, and at times happily abandoning the rigors of Aristotelian logic and the multiplication table. (Evans 1991: p. 299)

Further to this, the theory may actually hinder some road safety interventions by being quoted as a reason not to carry them out as any adaptation of behaviour caused by safety features is limited (Grayson 1996).

If the idea that people as individuals, and a collective, do not adjust their behaviour in response to feedback concerning current road safety statistics then another method of describing the relationship of risk to road use behaviour is needed. Fuller (2005) describes such a theory, one where drivers use proximal task difficulty rather than the aggregated level of system-wide risk to moderate their behaviour.

Task Difficulty Homeostasis

Risk Homeostasis Theory within the road safety environment argues that drivers and riders drive or ride in such away that they keep their level of risk at a target level. Task Difficulty Homeostasis, as suggested by Fuller (2005), purports that rather than subconsciously trying to keep statistical risk at a target level (Näätänen and Summala 1976; Wilde 1982) it is task difficulty that is maintained at a target level. In the Task Difficulty Homeostasis Model, task difficulty is the 'dynamic interface between the demands of the driving or riding task and the capability of the driver'. As seen from Figure 3.1, the model suggests that while a driver's or rider's task demand is lower than their capability then their driving or riding is in control; however, when task demand exceeds capability loss of control results, culminating in either 'a lucky escape' or a collision. Sometimes 'the lucky escape' is facilitated by other road users who manage to take actions that avoid a collision, such as swerving or performing an emergency stop. As task difficulty increases or capability decreases it would be expected that there would be a degradation of performance rather than a sudden loss of control (Wickens and Hollands 2000) with lower priority tasks such as, say, checking mirrors being neglected. As task difficulty further exceeds capability then more important tasks may not be carried out, such as proper forward observation. If task difficulty is increased to an extreme level then the rider or driver may only have enough attention and capability to focus on the road immediately in front, a form of tunnel vision.

Within the model of task difficulty shown in Figure 3.1 there is a feedback loop with a comparator at its centre, and this loop is shown in bold within the diagram (Fuller, Bates, Gormley, Hannigan, Stradling, Broughton, Kinnear and O'Dolan 2006). The comparator carries out the key function that allows adjustment of task demand by varying the vehicle speed. There are two controlling inputs to the comparator, acceptable task difficulty/risk and perceived task difficulty/risk. Perceived task difficulty is shown as a function between perceived task demand and perceived capability; therefore how a rider or driver views their own capability will affect the way that they ride or drive regardless of their actual capability.

One noticeable element of the model is that perceived risk and risk threshold are included as inputs to the comparator. However, this is not objective risk or statistical risk (Grayson, Maycock, Groeger, Hammond and Field 2003) as used in the Risk Homeostasis Model but rather this is perceived, or felt, risk. So rather than being based upon the rider's or driver's own calculation of the estimation of crash probability, this felt risk is an emotional response to danger (Summala 1986).

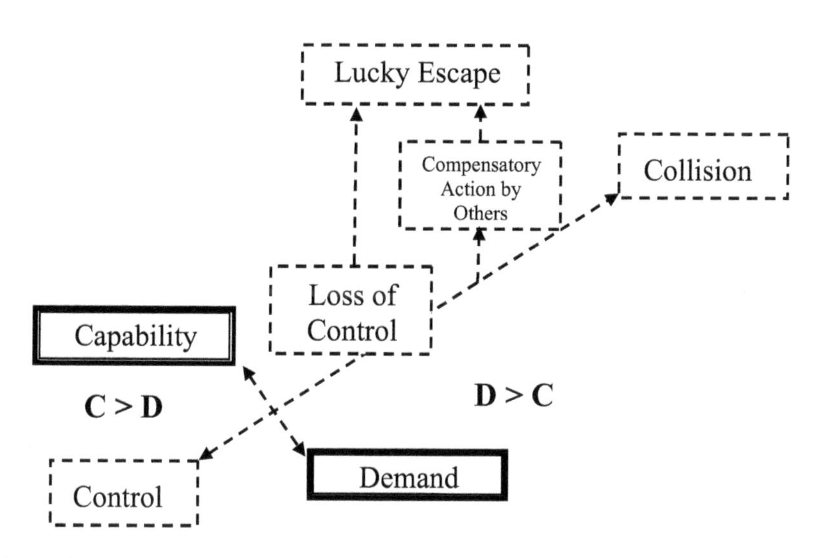

Figure 3.1 Outcomes of the dynamic interface between task demand and capability (Fuller, 2005: 464)

One of the other concepts of the Task Homeostasis Model is how close a rider or driver wants to position their acceptable task difficulty level to the out-of-control point where capability exceeds demand (C>D); that is, what safety margin do they desire to drive or ride with?

Figure 3.2 also contains elements that affect items within the feedback loop. The range of acceptable task difficulty is related to the amount of effort a rider or driver is willing to put into the task. Driving or riding goals are an essential element of setting the desired safety margin. For example, a person out for a nice casual drive to experience the countryside may have a higher safety margin than the same person who is late for a very important meeting.

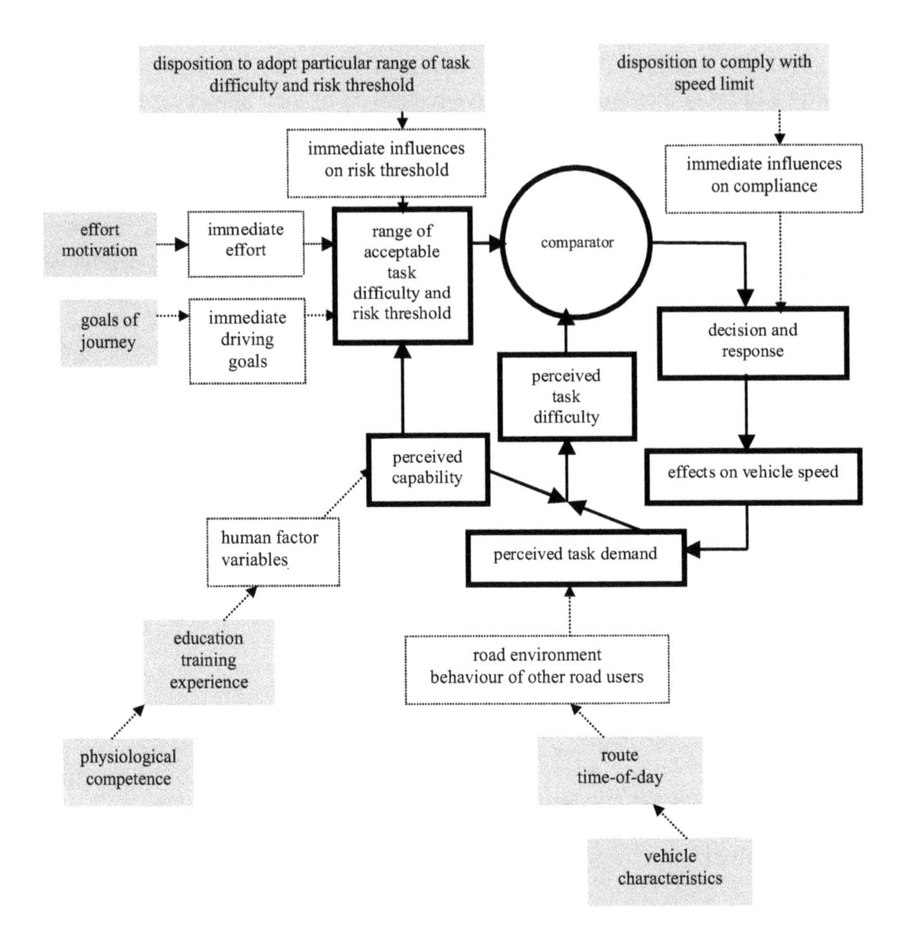

Figure 3.2 Representation of the process of task-difficulty homeostasis, distinguishing between proximal (clear boxes) and distal (grey boxes) determinants and influences on compliance

A person's temperament and the immediate local environment may also affect the level of the safety margin, for example, some riders may have a disposition to react angrily when they perceive that another road user has committed a violation against them and therefore they may ride with a lower safety margin until they have calmed down (Fessler, Pillsworth and Flamson 2004). Peers may also have an effect on how a rider might ride if a rider is susceptible to peer pressure. Data collected via a focus group noted that one rider when discussing how some ride said (Hannigan et al. 2008):

> ...You know the kind into speeding, at the top that's going to be the kind of weekend warrior types with the one-piece leathers and the state-of-the art sports bike. And the pressure is on them cause they've got a point to prove.

A rider's perceived capability is determined by how disposed they are to the riding task, training and other human factors. Perceived task demand is decided by two factors: the speed of the vehicle and the road environment itself. This is reflected in the comments of a rider (Hannigan et al, 2008):

> If it's in the summer, I may well ride faster than I would in the winter. In the winter time, certainly when it's getting round about the freezing, I will be concerned about the possibility of ice and slowing down.

As mentioned earlier, speed is an important factor in determining the level of task difficulty. The faster a rider goes the less time they have to process and react to hazards and therefore the whole riding task becomes more difficult. One rider expressed it (Hannigan et al. 2008):

> And again it was on the motorway, nobody else about, did it (high speed) for a couple of minutes, stopped whenever there was anything looking like it was getting too close. Just a bit too much sensory input for me, and a little bit too quick, even though feels like an empty road, it doesn't feel comfy.

What affects a rider's speed is of high importance. The output of the comparator is the main influence; however, a rider may also be constrained by the legal speed limit. For some riders keeping to a speed limit may be a natural thing that they always strive to do, while for others their speed compliance may only be achieved when they are being observed by the police or under the eye of a speed camera. The effect that fear of prosecution has upon rider speed is demonstrated by one rider who said (Hannigan et al. 2008):

> I think there was a comment earlier about losing your license, I mean, I'm sure most of us rely on our car licenses for our jobs, you know, so need our license for our job, and that is always kinda in the back of your mind as well, I think there is today that there are more opportunities to be caught for speeding than the old

days, and it would not take too many of …. those events for you to lose your license now. So as for me that is always a factor in the back of my mind.

Tasks

Panou, Bekiaris and Papakostopoulos (2005) derived eight driving tasks that combine to form the total task demand. Stradling and Anable (2007) expanded this to arrive at the 10 components, shown in Table 3.1.

Generally, for PTW users the task demand is higher compared to that of car drivers due to riding being a more complex task (Mannering and Grodsky 1995). Task demand is not only governed by the task of driving or riding but it is the total demand for all tasks being carried out. These extra tasks, such as trying to locate a particular turning, programming a satellite navigation system or using a mobile phone, can push up the total task demand beyond capability and hence place the road user at risk. Much of the research in the area of dual tasking has

Table 3.1 Ten components of the driving task (1–8 from Panou et al. 2005, 9 and 10 from Stradling and Anable 2007)

Task	Description
Strategic levels	Activity choice, mode and departure time choice. Discern route alternatives and travel time
Navigation tasks	Find and follow chosen or changed route; identify and use landmarks and other cues
Road tasks	Choose and keep correct position on road
Traffic tasks	Maintain mobility ('making progress') while avoiding collisions
Rule tasks	Obey rules, regulations, signs and signals
Handling tasks	Use in-car controls correctly and appropriately
Secondary tasks	Use in-car equipment such as cruise control, climate control, radio and mobile telephone without distracting from performance on primary tasks
Speed task	Maintain a speed appropriate to the conditions
Mood management task	Maintain driver subjective well-being, avoiding boredom and anxiety
Capability maintenance task	Avoid compromising driver capability with alcohol or other drugs (both illegal and prescription), fatigue or distraction

been concerned with mobile phone use by drivers (Haigney, Taylor and Westerman 2000; Laberge-Nadeau, Maag, Bellavance, Lapierre, Desjardins, Messier and Saidi 2003; Lamble, Kauranen, Laakso and Summala 1999).

Given the similarities of car driving and PTW riding, the 10 components of the driving task can be applied to riders. However, within these components, there are some important differences. For example, there are limited secondary tasks while riding compared to driving. Car drivers are enclosed in metal boxes that give opportunity for a plethora of secondary tasks, such as tuning the radio, programming the satellite navigation equipment, adjusting the heating controls, smoking a cigarette and occasionally even extreme and illegal activities such as using a mobile phone or shaving (BBC 2007g; Haigney, Taylor and Westerman 2000; Laberge-Nadeau et al. 2003; Townsend 2006). Some PTWs are now being fitted with satellite navigation equipment that has been adapted for rider use (Global Positioning Systems 2006); however, the majority of secondary tasks for riders are different from drivers. Tasks might include the adjustment of the helmet visor to demist it, or the acknowledgement of other riders. Therefore a PTW riding task list, shown in Table 3.2, was developed from these driving tasks (Broughton 2007).

For driving, 'avoiding collisions' is included in the 'traffic' tasks. However, as PTWs are vulnerable road users (BBC 2003; DfT 2006a; RoSPA 2001), and adverse interaction with road hazards are more likely to be serious or fatal (Clarke, Ward, Bartle and Truman 2004), hazard perception for riders is a very high level task. Therefore hazard perception as a major task has been included separately.

Hazard perception also interacts with many of the other tasks; for example, the road task of road positioning is partly defined by the perceived hazards presented by the road, such as over-banding and metal drain covers (Institute of Highway Incorporated Engineers 2005). The speed task is also dependent upon hazard perception as selecting the correct speed is partly hazard related, as well as being related to mood management, for example, a rider may ride faster if angry (Hannigan et al. 2008). The inclusion of hazard perception as a task brings the task components up to 11. The modified components for riders are listed in Table 3.2. The tasks in this list can be either proximal to the riding activity or distal. Task such as handling and speed are proximal as these are a direct response to the environment that the ride is being undertaken in. The strategic task is more distal, although decisions such as the route taken will have an effect on the proximal tasks.

Task Difficulty Homeostasis and 'The Spike'

Earlier in this chapter when discussing the theory of Risk Homeostasis, the idea was reported that if a large metal spike was fitted into the centre of the steering wheel then people would drive slower and this was attributed to Risk Homeostasis. Most people would agree that if a spike was fitted then people would drive slower,

Table 3.2 Eleven components of the riding task, adapted from Panou et al. 2005 and Stradling and Anable 2007

Task	Description
Strategic levels	Activity choice (functional and/or expressive) Departure time choice Discern route alternatives and travel time
Navigation tasks	Find and follow chosen or changed route; identify and use landmarks and other cues
Hazard perception	Detection of hazards
Road tasks	Choose and keep correct position on road, road position may be modified by road surface quality hazards
Traffic tasks	Maintain mobility ('making progress') while avoiding collisions (reaction to hazards)
Rule tasks	Obey rules, regulations, signs and signals
Handling tasks	Use PTW controls correctly and appropriately Interaction of PTW and rider (leaning at corners, etc.)
Secondary tasks	Keeping visor clean/demisted Acknowledgement of other riders Using satellite navigation equipment
Speed task	Maintain a speed appropriate to the conditions; speed will be modified by hazard perception
Mood management task	Maintain rider subjective well-being, avoiding boredom and anxiety
Capability maintenance task	Avoid compromising rider capability with alcohol or other drugs (both illegal and prescription), fatigue or distraction

so how can this be explained using Fuller's Task Difficulty based theory, after all the actual mechanics of the driver are not changed by the spike.

One of the comparator inputs within Fuller's model is the range of acceptable task difficulty/risk threshold. Two of the elements that control this are 'goals of the journey' (immediate driving goals) and 'immediate influences on risk threshold'. Both of these elements will be prejudiced by an immediate threat such as a metal spike. One of the goals of driving is to arrive safely and the obvious presence of a spike will influence this in the mind of the driver. The range of acceptable task demand will also be affected by the immediate threat, and increase in risk, posed by the spike.

Task demand may also be affected by the spike as the driver, being more than aware that it is there, will most likely spend some effort concentrating upon it, therefore adding an extra task to the others being undertaking in the driving activity.

Another difference between the theories is that Risk Homeostasis is based upon statistical risk while Task Homeostasis employs perceived risk, and having a metal spike sat a few inches from your chest would make a driver perceive the risk of the activity to be much higher.

The Somatic Marker

Damasio (2003) suggests that emotional signals from the local environment can influence the decision-making process. These somatic markers focus attention on the cause of the somatic stimuli. This is described by Damasio (2003) as emotions evaluating:

> … the environment within and around the organism, and respond accordingly and adaptively.

He goes on to state that these markers:

> … mark options and outcomes with a positive or negative signal that narrows the decision-space and increases the probability that the action will conform to past experience.

It is through experience that the markers induce a reaction with those experiences that create a probable negative result creating an attention grabbing emotional stimuli. The Task Homeostasis Model relies on a feeling that task demand is getting close to, or exceeding, capability and the idea of somatic markers could be the mechanism used (Fuller 2008; Kinnear, Stradling and McVey 2008). Riders and drivers may use the level of anxiety that they feel to set their desired speed (Taylor 1964) and maintain a constant safety margin. Within Fuller's (2005) experimental work on Task Homeostasis he discovered that the feelings of risk, task difficulty ratings and driving speed had a very high level of correlation, and his data is plotted in Figure 3.3. This correlation is what would be expected as anxiety, felt as risk, increases due to the higher vehicle speed.

The idea of Task Homeostasis, linked with somatic markers, demonstrates that the riding and driving decision-making process is not a purely cognitive skill, but rather a combination of thinking and feeling.

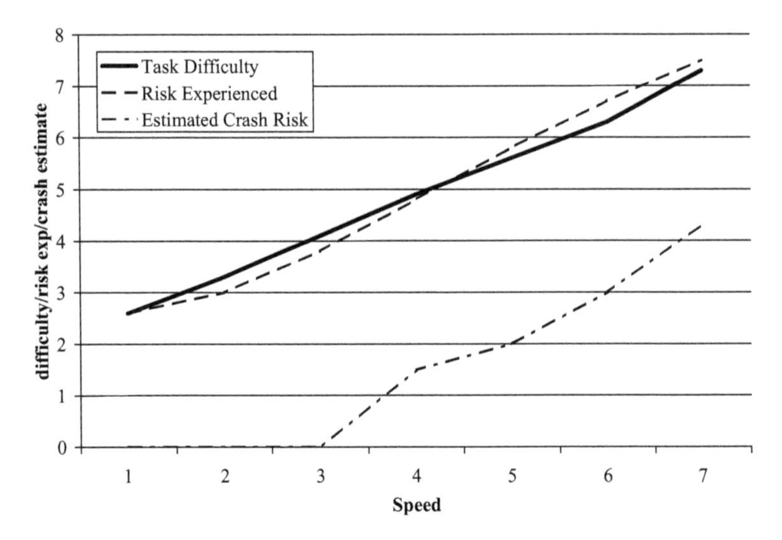

Figure 3.3 Ratings of task difficulty, estimates of crash frequency and ratings of risk experience (Data extracted from Fuller 2005: 469)

Motorcycle Crashes and Task Difficulty

Speed is a major way of controlling task difficulty with an increase in speed making the riding task more difficult, therefore it is not surprising that riders tend to ride slower when task difficulty is high. However, where task demand exceeds capability (D>C) then an out-of-control situation exists that will result in either a lucky escape or a crash for the motorcyclist as depicted in Fuller's model (Figure 3.1). As crashes result from this D>C situation then it should be possible to explain motorcycle crashes in terms of task difficulty (Broughton 2008a). To illustrate this two common types of crashes that involve bikes will be considered: loss of control on bends and crashes while overtaking.

Clarke et al. reported that 12 per cent of KSI crashes resulted from loss of control on a bend (Clarke, Ward, Bartle and Truman 2004) while a Scottish study that analysed motorcycle crashes between 1992 and 2002 found that 20 per cent occurred on bends (Sexton, Fletcher and Hamilton 2004). So what is it that makes bends so problematic? Part of the problem is that riders may be riding with a small safety margin and therefore it only takes a small error in how the rider sets himself up for the corner for an 'out-of-control' situation to occur. This can be understood when a bend is considered at the end of a straight piece of road. Riding along in a straight line does not require much capability and the task difficulty is low. However, that will change on the approach to the bend when the task difficulty will quickly increase. If a rider is not experienced enough to react quickly, then he may find that the task difficulty is beyond his capability and he is out of control.

Task difficulty can also unexpectedly change, for example, by a sudden reduction in traction between tyres and the road surface due to a diesel spillage or similar problem, or where a rider's capability is reduced by being distracted by occurrences such as an animal running into the road.

Around 14 per cent of crashes occur while the bike is overtaking (Sexton, Fletcher and Hamilton 2004). The task difficulty of riding in a straight line is relatively low, however, to overtake, the rider will most likely accelerate and any rapid increase in speed will result in a corresponding rapid increase in task demand. This may be a particular problem as bikes can accelerate significantly quicker than most cars. For example, a Ford Focus ST car can accelerate from 0 to 60 mph in 6.8 seconds, while a BMW F800s motorbike can do 0 to 60 mph in 3.5 seconds; therefore task difficulty can rapidly increase to a point where task demand exceeds rider capability.

Given the vulnerable nature of riding a PTW with its limited protection (ACEM 2004; DfT 2005) making any such lapse that puts the rider into a D>C condition is more likely to result in serious injury or death than for a car driver.

Summary

This chapter has examined two of the key theories on risk used in road safety and applied them in relation to PTWs. Task Homeostasis Theory would seem to offer a more plausible explanation of the way that riders face challenges presented to them while riding. It would seem to offer an explanation for the way that PTW crashes may happen when task difficulty exceeds task capability. However, while it may offer explanations for aspects of riders' behaviour while riding, it does not address the more fundamental issue of why riders choose to use PTWs given the additional risk of death and serious injury faced through this choice.

Chapter 4
Risk and Sensation Seeking

The connection between sensation seeking, motorcycle use and rider crash risk is one that has been much discussed (Mannering and Grodsky 1995; Sexton, Hamilton, Baughan, Stradling and Broughton 2006). Part of this argument stems from the fact that riders have a higher risk of being seriously injured or killed than four-wheel road users (DfT 2007c; RoSPA 2001); some believe that it is this increased risk that acts as an activity attractor. While the previous chapter examined theories of risk, there was little discussion of what risk is. This chapter explores issues of definition before examining risk taking behaviour and the related concept of sensation seeking.

Risk

Risk can be defined in a number ways, for example:

> The chance of injury, damage, or loss. Therefore, to put oneself 'at risk' means to participate either voluntarily or involuntarily in an activity or activities that could lead to injury, damage, or loss. (Webster's 1979)

> The quantitative or qualitative expression of possible loss that considers both the probability that a hazard will cause harm and the consequences of that event. (Lawrence Livermore National Laboratory 2005)

The latter definition suggests that risk is a function of probability of a hazard causing harm and the level of harm or consequences:

Risk = f(Probability, Consequences)

If risk is a function of the probability of a hazard causing harm or consequences, then how can a hazard be defined? A hazard can be defined as:

> A source of danger (that is, material, energy source, or operation) with the potential to cause illness, injury, or death to personnel or damage to a facility or the environment. (Lawrence Livermore National Laboratory 2005)

Therefore hazards and risks are two different, albeit related, items. A hazard can be distinguished from a risk as being a specific danger; this is expressed by Sharp:

> Hazards are defined in absolute terms (for example, cliff faces, avalanche prone slopes, fast moving water, electricity, sharp knives) … (Sharp 2001: p. 10)

So a hazard has the potential to cause harm while risk on the other hand is the likelihood of harm occurring and is usually qualified by some statement of the potential severity of the harm. Risk is particular to the person that is confronting the hazard and also the context in which the hazard exists at that instant in time. For example, the risk of riding on a wet road by a person with limited experience would be greater than the risk on the same road being used in the dry by a highly trained police rider.

The other implication of this is thus that what might be risky for one rider may not be risky for another. Experience is a major factor in this as the more often a hazard is encountered the lower the risk as the rider will have a better understanding of the hazard and how to negotiate it safely (Clarke, Ward, Truman and Bartle 2003; RoSPA 2006). Risk is also dynamic as it changes as circumstances change. It is a combination of the probability of a hazard triggering an event and the likely severity of consequences if that event occurred. The chance of a hazard triggered event happening is related to exposure time, that is, the more time that one is exposed to a risk the greater the chance that one will be affected by it. This will be mitigated in part by the experience gained by this exposure giving a greater awareness of the hazard, how to avoid it or how to react when faced with it. For example, more experienced riders will be more aware of the danger of a diesel spillage and where it is likely to occur but if they ride regularly they would almost inevitably find themselves faced with this hazard. When this happens they will be less likely to brake hard and hence be less likely to crash. The level of risk can be therefore be reduced by:

- reducing the probability of an incident;
- reducing the exposure time;
- ameliorating the consequences.

Some geographic areas have sent out messages that may be interpreted by some as being anti-bike; for example, Mr Wolfendale of North Wales Police asked:

> Is it now time to banish motorcycles completely from our national parks? (BBC 2006)

This message, and others like it, may make some riders think twice about riding in a specific area, thus reducing the exposure and the number of motorcycle crashes within that locale. However, this is likely to simply have a displacement effect causing an increase in crashes in adjacent areas.

As discussed in Chapter 2, the level of severity of a crash is related to speed as the amount of energy expended in a collision is proportional to the square of the speed (Aarts and Van Shagen 2006), and therefore a doubling of speed at impact will result in four times the energy being absorbed by immovable tree, dry stone wall, hard metal, soft flesh or brittle bone. Just this fact alone demonstrates that an increase in speed causes an increase in risk.

Levels of acceptable risk also vary between people and situations. Related to risk acceptance is the amount that can be gained by undertaking a risky activity with acceptance being a trade-off between perceived risk and perceived gain (Coombs, Donnell and Kirk 1978). Personality factors, coupled with motives, may also be a deciding factor on whether people take part in low- or high-risk activities, such as sports (Diehm and Armatas 2004) and other recreations.

Statistical risk is very hard to calculate and therefore some form of estimation is used when people refer to risk. This means that generally when risk is discussed it is perceived risk rather than statistical risk that is being referred to. This perceived risk is not only based upon cognitive processing of risk-related information but also on emotional reactions to circumstances (Kinnear, Stradling and McVey 2008). The feeling of risk therefore often has very little to do with real, or statistically accurate, levels of risk. With the correct information it is possible to accurately determine what the statistical risks of activities are, but typically in the real world only imperfect information is available as a decision has to be made quickly. When information is inaccurate or a decision has to be made under time constraints then a heuristic method may be used; a heuristic is a basic rule of thumb. With heuristics when an event occurs a predefined action is taken, this is sometimes referred to as 'fast and frugal' (Gigerenzer and Todd 1999). For riders, this initially results in habitual actions being taken when a threatening hazard is detected and then, if time allows, more thought through actions can be taken.

When considering the consequences it is important that not only physical harm is considered as a loss; losses can also be financial, social or temporal (Rohrmann 2002). Walker (2007) quoted seven types of risk, based on Rohrmann (Table 4.1).

For a rider, financial losses can occur in more than one way. For example, if a rider has a crash there is likely to be some financial cost in the form of having to pay insurance excess and an increase in premiums in the following years. There would also be a financial penalty if a rider is prosecuted for a road traffic offence in the form of a fine, and maybe a concomitant increase in insurance premiums. A rider may suffer social losses by losing the respect of his riding cohort, having a crash and looking like a poor rider. A crash may have a functional risk as well as the potential of generating psychological risk. A rider may also feel that there is a risk to their ego if they are out-ridden by other riders. A rider may lose riding, as well as working and leisure time, if an injury is suffered after a crash. A crash may also reduce riding time, as the rider may not have a bike to ride; similarly riding time could also be curtailed if a rider is banned from riding after prosecution for a road traffic offence.

Table 4.1 Risk types, adapted from Walker 2007 and Rohrmann 2002

Type of risk	Meaning
Functional risk	Unsatisfactory performance outcomes
Financial risk	Monetary loss, unexpected costs
Temporal risk	Wasting time, consequences of delays
Physical risk	Personal injury or damage to possessions
Psychological risk	Personal fears and emotions
Social risk	How others will think and react
Sensory risk	Unwanted impacts on any of the five senses

Risk Takers

We all do risk assessments as part of everyday life, most of which are typically fast and frugal such as when we cross the road. There is little time spent on assessing the actual risk, that is, how fast a car is approaching the spot or the statistical likelihood that a car will pull out from a junction or parking place. The decision is made to cross from a cursory appraisal based on experience. How good is this assumption of risk? Often it is not good as it is clouded by other factors (Freudenburg 1998):

- familiarity with the action;
- the perceived danger, which is often incorrectly estimated due to lack of knowledge;
- how much we want to carry out an activity, that is the reward.

The idea of how risk is viewed was expressed by Lord Rothschild:

> There is no point in getting into a panic about the risks of life until you have compared the risks which worry you with those that don't, but perhaps should. (Rothschild 1979)

Perceived risk plays an important part in road safety as 'decision skill' within driving or riding is an area where most errors occur and therefore is the main underlying factor of road crashes (Colbourn 1978). The decisions that drivers make are, to a great extent, down to how risky they perceive the situation, meaning that crashes can and do occur due to drivers and riders underestimating the risk of the situation they face. Colbourn also explains that other tasks or motivationally-based variables may have an additional effect on how a person perceives the risk of a specific situation. Therefore the perceived risk in a given situation may be different for different individuals (Rohrmann 2002).

It is widely accepted that people have a general orientation towards risk; an attitude towards taking risks (risk propensity) or towards avoiding risks (risk aversion). Risk propensity and risk aversion are attitudes, not behaviours; that is they are cognitions that precede behaviour (Rohrmann 2002). What one person enjoys and seeks may be highly aversive to another, for example, some may enjoy the experience and intensity of a horror film or roller coaster ride while others may enjoy light classical music; some may enjoy playing contact sports with a high risk of injury such as rugby while for others a low risk sport such as bowls suits their risk orientation better (Broughton and Stradling 2005). Some may indulge in 'risky' activities as a means to satisfy their arousal needs; some may be 'attracted to', rather than 'scared away from', a risky situation (Lupton 1999). Another group that may be in the 'take higher risk' bracket are those who are not very good at gauging risk, their risk assessment not well calibrated; for instance their perception of risk may be considerably lower than the real risk that the situation poses. People in this group may not be in the 'risk propensity' class but may still undertake risky activities that may lead to similar behaviours to those who are risk averse, but for different reasons.

Risk taking attitudes are an important factor in road safety as those who are involved in crashes are more likely to be those who take higher risks (Iversen 2004; Turner and McClure 2004; Turner, McClure and Pirozzo 2004). Risk propensity towards driving, that is risky driving behaviour, among young drivers is predominantly a male activity and it is also mainly males who go on to be risky drivers later in life. In general, women show high risk taking behaviour less often than men (Siegrist, Cventkovich and Gutscher 2002). This could be partly due to attitudes to risky activities being influenced by the social ideas of masculine and feminine identity (Lupton 1999). Some people, denoted as risk seekers, deliberately take risks; they may do this for pleasure or maybe to rebel against the self-control and self-regulation that society places upon them (Lupton 1999). Fessler, Pillsworth and Flamson (2004) found that anger increases risk taking in males, while disgust reduces risk taking in females. This shows that a person's emotional state can influence their risk taking behaviours and that emotional state reactions differ for males and females.

What is the reaction when the risk of an activity is reduced and the activity made safer? Peltzeman (1975) theorised that people would compensate for improvements in car transport safety by driving in a more risky manner; this has been called the theory of risk compensation. This also applies to those who ride motorcycles; Chesham, Rutter and Quine (1993) said:

> A real reduction in motorcycling accidents can be achieved only by changing the level of risk found acceptable by riders when operating their machines. (Chesham, Rutter and Quine 1993: p. 425)

What happens if the risk, or perceived risk, of an activity increases? Noland (1994) looked at this with regard to modes of transport and found that if the perception of risk increased for a mode of transport, such as the bicycle, then the probability of that mode being used for commuting decreased. If an improvement is made so that a mode of transport is made safer such as the provision of cycle lanes, then more people may use that mode of transport. Therefore the reduction in fatalities may not be proportional to the reduction in risk due to the increased level of exposure; that is, the number of fatalities may not decrease in absolute terms but might do in terms of the numbers of those using that mode of transport. Another suggestion of Noland's research is that people will choose a route to commute that they feel minimises their risk. However, those who have a leaning towards risk propensity, or who are risk seekers, may be less affected by these changes and may even react the opposite way by not participating in activities that are made safer, migrating to more dangerous ones.

Sensation Seeking

One of the factors that is often associated with motorcycling and the risks involved is sensation seeking. Sensation seeking is a personality trait that has been linked to decision-making concerning risky actions (Zuckerman 1979; Zuckerman 1991). Zuckerman describes sensation seeking as:

> ... the need for varied, novel, and complex sensation and experiences, and the willingness to take physical and social risks for the sake of such experiences. (Zuckerman 1979: p. 10)

Sometimes where a high level of sensation seeking would be expected it is not found. For instance it may be believed that people who take part in contact sports would be high sensation seekers, yet O'Sullivan, Zuckerman and Kraft (1998) found that this was not the case, rather he stated that sensation seeking is a feature of:

> ... participants in high risk sports offering unusual sensation and personal challenges.

A study on risk taking and sensation seeking showed that risk takers seem to be higher in sensation seeking then other members of the general population (Fischer and Smith 2004; Horvath and Zuckerman 1993). Also, drivers who have a higher sensation seeking score on the Zuckerman Sensation Seeking Scale were more likely to be involved in a crash and drive in a more risky fashion (Heino, van der Molen and Wilde 1996). The Zuckerman Sensation Seeking Scale (SSS-V) (Zuckerman 1983) is commonly used to assess four aspects of sensation seeking:

- Thrill and Adventure Seeking (TAS)
- Experience Seeking (ES)
- Dis-inhibition (DIS)
- Boredom Susceptibility (BS).

Generally, participants of high-risk sports have significantly higher scores than the control group on TAS, ES and Total Sensation Seeking (TotSS) (Freixanet 1991).

Another, and simplified, scale for measuring sensation seeking is Arnett's Inventory of Sensation Seeking (AISS) (Arnett 1994). This scale measures just two aspects of sensation seeking:

- intensity;
- novelty.

Intensity is the desire for intensity of stimulation to the senses while novelty is to do with the person being open to the experience.

Arnett's Inventory of Sensation Seeking (AISS) is used here in preference to Zuckerman's version as it is more contemporary that Zuckerman's version. The AISS uses 20 descriptive statements that the subject grades on a four-point Likert Scale where one is 'does not describe me at all' and four is 'describes me very well'. Therefore the total sensation scale ranges from 20 (low sensation seeking) to 80 (high sensation seeking).

The AISS assessment system does not include items that are age-related, would discriminate those who lack physical strength or would not be involved in norm-breaking behaviour. Another advantage of the AISS over the SSS-V is that the AISS questionnaire is shorter and this shorter questionnaire may aid in getting a better response rate.

The two 10 item subscales (intensity and novelty) within the AISS have scores ranging from 10 to 40. Arnett stated that:

> Sensation seeking is not only a potential for taking risks, but is more generally a quality of seeking intensity and novelty in sensory experiences, which may be expressed in multiple areas of a person's life. (Arnett 1994: p. 290)

Risk taking behaviour plays a large role in the contribution to car/PTW crashes that result in injury. Turner and McClure (2004) hypothesised that people who have a high-risk acceptance level perceive risk differently from those that do not. This leads them to drive or ride in a more risky manner, which in turn leads to them being involved in more crashes. In their study only 4.6 per cent of people were defined as having a high-risk threshold, yet this group were involved in 25.3 per cent of the crashes involving injury. From this it was concluded that if 'high-risk acceptance' could be removed then the injury crash rate would be drastically reduced.

Conversely, Turner and McClure found that those who had a high-thrill seeking behaviour did not have an increase in injury from a crash and they suggested that thrill seekers are less likely to be injured as they are better equipped to deal with risky activities (Turner and McClure 2004). Heino, van der Molen and Wilde (1996) reported that sensation seekers followed cars closer than those who had risk aversion, which is not unexpected. Driver and rider behaviour is related to sensation seeking with those who have a higher sensation seeking score being more likely to commit traffic violations and errors (Ball and Zuckerman 1992).

Are there other factors that are involved with sensation seeking that can help to explain the relationship of crashes and risk taking? Fischer and Smith (2004) suggested that impulsiveness should be considered with sensation seekers. They found that individuals who experience negative life outcomes were more impulsive (less self-control and constraint) than those who do not. A lack of deliberation, or being impulsive, can be described:

> ...as a failure to plan ahead, or acting without thinking. (Fischer and Smith 2004: 528)

One can be a deliberate sensation seeker and as such, one is less likely to suffer negative results compared to an impulsive sensation seeker. Those sensation seekers who take part in risky sports who are from the deliberate sensation seekers sub-group are more likely to be successful and to plan ahead with safety measures than those from the impulsive sensation seekers sub-group. While there is a positive relationship between those who take part in high-risk sports and sensation seeking, Zuckerman (1992) emphasised that risk taking is not an essential motivation for sensation seeking behaviour. Ben-Ari, Mikulincer and Iram's (2004) research into young drivers' attitudes supports this as they mostly have a disregard for negative consequences because young drivers do not take the time to deliberate upon them.

In investigating the sensation seeking levels of rider, data has been taken from two sources. Source one was collected specifically for this publication with the riders being recruited online using forums, and via motorcycling clubs with the snowballing technique used to reach other riders, that is encouraging the respondents to get others (friends, relatives, colleagues, acquaintances) to participate (Coomber 1997; Hewson, Yule, Laurent and Vogel 2003). Source two data is taken from the Risk and Motorcyclists in Scotland report for the Scottish Executive (Sexton et al. 2006). The mean scores are summarised in Table 4.2.

Both of these studies show that riders have a slightly higher level of novelty than intensity. This may seem to be counter-intuitive as it may be theorised that it is the intensity of the riding that attracts riders. However, these results suggest that the seeking of new experiences is at least as important and that this personality trait is one of the keys to attracting potential riders into biking. This is supported by comparable research into sports where Zuckerman concluded that differences between participants in high- and low-risk sports are due in part to the need for some novel sensations and new experiences (Zuckerman 1994).

AISS data was also collected from non-riding car drivers for this publication to allow a comparison between drivers and riders. Table 4.3 shows this comparison.

There is no difference in intensity between drivers and riders; both groups tend to seek the same level of stimulation to the senses. However, there is a statistically significant difference between riders and drivers in the novelty sub-scale, with riders being more likely to seek new experiences. It is the disparity in this sub-scale that influences the difference of the overall combined results. Those who have a higher level of sensation seeking generally perform better at attention tasks (Ball and Zuckerman 1992) and as motorcyclists are more vulnerable than drivers, this better attention may be deployed in hazard perception and thus aid in protecting riders.

Those who take part in sensation seeking behaviour do not necessarily do so because they want to take part in risk taking activities. Rather risk taking is the result of pursuing activities that generate novelty, change and excitement. Many of the activities that sensation seekers undertake are not risky but involve stimulation such as rock music, watching horror films and visiting exotic places.

Table 4.2 Summary of AISS data for riders

	Source 1	Source 2	Combined
n	63	335	398
Novelty	27.2	26.6	26.7
Intensity	26.9	23.7	24.2
Combined	54.1	50.3	50.9

Table 4.3 A comparison of mean AISS between drivers and riders

	Rider mean	Driver mean	Significance
Intensity	26.7	26.6	NS
Novelty	24.2	23.3	$t(459) = 3.216, p = 0.002$
Combined	50.9	49.9	$t(459) = 1.936, p = 0.058$

Those who undertake risk taking sports such as mountaineering do not do it because they enjoy the risk of serious injury or death but because they feel that they can control the risks with the skills that they possess (Zuckerman 1983). The risks that are taken may trigger emotional physiological response such as an adrenaline rush, however, these are not viewed as life threatening because of the sportsman's belief that they are in control (O'Sullivan, Zuckerman and Kraft 1998). This type of participant gains enjoyment and satisfaction in the exercise of control in potentially dangerous environments.

Summary

Riding PTWs for pleasure is a risky activity and it has generally been assumed that those who ride for pleasure do so because it is a risky activity. If this were the case, then it would be expected that riders would score highly on the intensity aspects of the AISS scale whereas analysis of rider data suggests that are seeking novelty of experience. Research into extreme/adventure sports such as mountaineering suggests that people undertaking such activities are enjoying the challenge of matching their skills to the challenge. Therefore, the sports person is finding satisfaction from exercising control over a potentially hazardous situation. It may be that PTW riders have similar motivations. The following chapter seeks to contextualise the current PTW situation by exploring the historical and cultural background of PTW riding in Western culture and examine the characteristics and profile of today's PTW riders.

Chapter 5
Who Rides?

Introduction

Social theorists such as Wright Mills maintain that 'history is the shank of social study' (Wright Mills 1959: p. 143) and that the psychology of man should be understood in its sociological and historical context. Therefore, although later chapters are primarily concerned with the beliefs, attitudes and behaviours of riders, and to a lesser extent non-riders, this chapter concentrates on the way that biking has developed in the United Kingdom and the way that it has been portrayed in the media to develop an understanding of 'biking' today.

While functionally the Powered Two-Wheeler (PTW) can be seen simply as a mode of transport, the motorbike is embroiled with cultural meaning and embedded in popular culture (Alt 1982). Motorbikes, their riders and associated meanings go beyond their functionality. In order to understand the cultural identity of bikes and bikers, the development of biking and its associated media image must be explored. Many of the current conceptions of bikers are based on portrayals in news and entertainment media. Such portrayals were often negative and misleading or based upon the behaviour of a small minority within the biking fraternity. Therefore this chapter seeks to explore the facts and mythology relating to PTWs and their riders.

The Developing Image of PTW Riders

When motorcycles and biking were in their infancy, those who rode were enthusiasts, engineers, people who enjoyed tinkering with bikes and eccentrics. This image is demonstrated with a quote taken from the early days of motorised transport by Donald Heather, Director of Norton:

> Most motorcyclists love to spend their Sunday mornings taking off the cylinder head and re-seating the valves. (Hopwood 1998)

However, the Second World War changed the nature of biking and the image of bikers. Many soldiers were affected by their wartime experiences and felt the need for a sense of identity and freedom, and some found this in biking. At this time the trend of racing from café to café started with riders dressing in leather jackets. Motorcycle gangs were forming and this was accompanied by bad press which in turn put many off from buying a bike and opting for the family friendly car instead (Quiñones 2006).

In 1954, less than 10 years after the war ended, the image of the biker as a criminal was established when Columbia/Tristar Studios released *The Wild One* starring Marlon Brando. This film did more than anything else to establish the negative image of bikers in modern culture (Dirks 2006b). The film was very loosely based on a real life story. On the weekend of 4 July 1947 about 4,000 people descended on the town of Hollister in California, many riding bikes. However, unlike in the film where the town was destroyed, in fact only a few arrests were made and these were mainly for drunkenness. This event was reported in the January 1951 issue of *Harpers Magazine* in an article entitled 'The Cyclists' Raid' (Rooney 1951). In 1954 another article reported, 'Nobody – except another cyclist – likes a man on a motorcycle' (Burton 1954). In 1969 the outlaw side of biking was further emphasised with the release of the film *Easy Rider* starring Peter Fonda, Dennis Hopper and Jack Nicholson (Dirks 2006a). This film modified the image of bikers in popular to be an outlaw criminal, living on the edge of society.

Various events and particularly the way that they have been reported by the media have also played their part in generating the 'outlaw' image of riders. The conflict between the 'Mods and Rockers' in 1964 is one such example (BBC 2007a). The Rockers based themselves on motorcycle gangs wearing black leather and riding motorbikes; they are often thought of by the public as the quintessential biker (Stuart 1987). During the Whitsun holiday weekend gangs of 'Mods and Rockers', by chance, met up in Brighton. This situation soon gave way to violence with the fighting lasting for two days and stretching from Brighton to Hastings; for this reason this incident is sometimes referred to as 'The Second Battle of Hastings'. The events of that weekend were further ingrained into the public psyche with the release of the 1979 film *Quadrophenia*, based on the events of that weekend.

How justified is this image of bikers and motorcyclists? Do they want to be outlaws living outside of society? If the image of riders that is held by society is based on misconceptions then it should be corrected; however, the way that certain groups are viewed cannot be changed easily and any changes that are made are often slow (Hogg and Abrams 1988).

Motorcycle Subculture – Community, Gangs, Clubs and Organisations

Biker or motorcyclist, what is the correct title? Within the biking community there are some that refer to themselves as bikers, others prefer the title of motorcyclists or rider. Which title is used often depends upon the rider themselves; however, the reaction from the public may be different depending on the title. One motorcyclist commented:

> I think the term 'biker' in itself draws negative associations to those who choose to ride motorcycles. (Broughton 2007: p. 419)

An example of this perception is give by a member of Cossack Owners Club:

> I have in the past booked campsites for our rallies, the owners (members of the public), did not mind hosting a motorcycle rally, but said that that they refuse bikers as they have had trouble with them in the past. Together we are two groups of motorcycle riders that are seen very differently by many members of the public. (Cossack Owners Club 2008)

Motorcycling, despite mainly being a solo activity, can also be very social with riders having a sense of belonging to a community. Many would have noticed how riders acknowledge each other as they ride, despite most likely having never met the other motorcyclist. People's identity, and that includes bikers, comes in part from the groups that they belong, or believe that they belong, to (Hogg and Abrams 1988).

Motorcycle Outlaw Gangs

When the phrase motorcycle club is mentioned, many people's thoughts go straight to gangs such as the 'Hell's Angels'. This has a lot to do with popular culture as discussed in the Introduction, however, most motorcycle club members bear little resemblance to this media generated image. Organisations that are loosely called 'outlaw gangs' such as the Hell's Angels are sometimes referred to as 'one percenters', a title that many of them wear with pride, even to the point of wearing a patch with '1%' on it as part of their insignia.

The 'one percenter' designation came into existence after the now infamous Hollister weekend that gave birth to the *Wild Ones* film. The American Motorcyclist Association (AMA) was asked to comment on the film. It has now been written into folk law that the AMA said that 99 per cent of riders were law abiding and therefore by deduction one per cent were outlaws. However, Tom Lindsay of the AMA said:

> We acknowledge that the term 'one-percenter' has long been (and likely will continue to be) attributed to the American Motorcyclist Association, but we've been unable to attribute its original use to an AMA official or published statement. (Dulaney 2005)

In truth, the number of riders who are members of these outlaw one percenter clubs make up significantly less than one per cent of the rider population. (Gutkind 2008)

The image of these biker gangs is further reinforced by media coverage of events that involve them. The events of 1998 are still remembered by many when, as part of a long running feud, there was a fight between the Outlaws and the Hell's Angels that left two dead (Guardian 1999). This story was resurrected by the BBC in 2007 after the murder of a Hell's Angel, Gerard Tobin, while riding on the M40 in Warwickshire, England. Reporting on the M40 incident, the BBC also commented that around 40 Hell's Angels were involved in ambushing members of the Outlaws in 1998 (BBC 2007f), nearly 10 years previously. Media coverage like this reinforces the 'Wild One' image of riders. While not condoning the behaviour of these gangs, the publicity of such stories seems out of proportion when compared to the number of deaths and serious injury caused by non-bike-related gangs.

Membership of these one percenter clubs is not for the faint hearted as members are expected to 'live the life' 24 hours a day; it is not a part-time activity. Potential new members, or prospects, are tested as to their suitability for joining the club. Within most organisations, they require a unanimous vote from existing members to become fully associated with the club. Each of these clubs has its own 'colours' that their members wear with these identifying the members of the club as well as identifying which section, or chapter, that they belong to (motorcycle.org.uk 2008).

While most people have heard of the Hell's Angels, they may not have heard of other outlaw gangs such as the Blue Angels, who originated in Glasgow, and the Bandidos who now claim to be gaining a foothold within the UK from their European base (BBC 2008b)

This image of outlaw bikers is ingrained within the minds of most people such that wherever gangs like the Hell's Angels go, they would expect trouble. For example, *Time* magazine reported on 26 Mar 1965 (*Time* magazine 2008), while discussing the Hell's Angles that:

> Their initiation rite, for example, demands that any new member bring a woman or girl (called a 'sheep') who is willing to submit to sexual intercourse with each member of the club. But their favorite activity seems to be terrorizing whole towns.

However, trouble and these gangs do not always go hand in hand, for example, over the Easter period in 2008 it was the 60th anniversary of the Hell's Angels and many expected trouble at the celebratory party. *Clutch and Chrome* (*Clutch and Chrome* 2008) magazine reported from the event that:

> A major concern was the party's effect on local worshipers observing the Easter weekend, but despite the reputation of the Hells Angels, two nights of celebrating the 60th year of the motorcycle gang's existence passed without any reported trouble.

Much has been written about the outlaw club culture – for example, see (Dulaney 2005; Sher and Marsden 2006; Veno 2003), however, the majority of riders are not part of this culture.

Motorcycle Clubs

The majority of motorcycle clubs are not part of the outlaw culture but rather based upon the commonality of riding. Clubs exist for many reasons and with different purposes. Some clubs, such as the Scottish Motorcycle Club, have the aim to organise group riding trips and describes itself as:

> …an independent non-profit-making Scottish club open to all who are interested in riding motorcycles responsibly for fun and social pleasure in the company of other members. (The Scottish Motorcycle Club 2006)

Organised rideouts are a key element of many clubs. 'Rideouts' are arranged runs where a group of riders get together to follow a pre-determined route organised by one, or more, of the group members.

Some clubs are designed for a specific membership cohort such as the Perthshire Ladies MCC while other clubs may only allow riders of specific bikes to join, for example, the BMW club; or aimed at vintage bikes, for example, Vintage Motor Cycle Club.

Despite the nature of most clubs the image that is conjured up by non-riders is still one of danger and hostility. This image though would seem to be at odds with one specific group of riders – Christian Bikers. The Christian Motorcycle Association (CMA UK) within the UK has it roots back to 1979 when advertisements were placed in national biking magazines enquiring if any Christian riders wanted to get together. The CMA (UK) is now part of a worldwide network of Christian biker associations and clubs; they have even published their own edition of the Bible.

The social aspect of motorcycle clubs should not be under-estimated (MCIA 2008b). Some clubs meet at regular intervals where the aim is not the riding of machines but the chance to socialise and discuss their riding activities with like-minded people.

The Internet and Rider Forums

Modern technology has given birth to rider internet forums and virtual motorcycle clubs. Sites like the UK Bike Forum (UKBF) allow riders to discuss virtually any subject, from how to set up the suspension of a bike to what happened in a soap opera. One member of the UKBF stated that:

> As a beginner to biking, it gives me the chance to ask questions without feeling like a numpty. People will give their honest opinions and there are a lot of people who are specialists in their areas which is really useful. I also get to add my opinion/expertise at times and help out others where I can, thus giving back to an ever-growing community.

Forums and virtual clubs can take many forms, such as for riders of specific makes or models, but they give the chance for riders to associate with a large number of riders, if only in cyberspace, although some of these virtual communities do organise meetings and 'rideouts'.

Motorcycle Organisations

Similar to clubs are the motorcycle organisations that promote or politically defend riding and riders' rights. The Motorcycle Industry Association (MCIA) is one of these organisations; this group represents the manufacturers and importers of bikes and biking equipment. Their aim is:

> To develop and sustain an environment whereby motorcycling can flourish. (MCIA 2007a)

They do this mainly by lobbying and discussing biking issues with the relevant authorities. They are also involved in road safety promotion and research (MCIA 2006a), organise the national ride to work day (MCIA 2008b) and also organise the annual National Motorcycle and Scooter Show that is held each year at the National Exhibition Centre (NEC) in Birmingham (MCIA 2008a).

There are two main rider rights organisations within the UK: the British Motorcycle Federation (BMF) and the Motorcycle Action Group (MAG). The BMF claims to have around 92,000 members (BMF 2008) and they state that:

> Our aim is simple – to promote and protect the interests of the road rider by representing the rider's interests where and when it matters.

MAG was formed in 1973 (MAG 2008) to protest against compulsory helmet use and has similar goals to the BMF, with MAG stating that it remains faithful to its:

> … core principles of freedom of choice and self-determination.
>
> In a world of increasing regulation and conformity motorcycling represents one of the last bastions of individuality and diversity. This freedom of expression and individuality is not a 'given'; it has to be protected and fought for.

Both of these organisations are not only involved in political lobbying, but they also run motorcycle events, such as the Edinburgh based 'Ag Ol Anns An Achaidh Rally' organised by the Edinburgh section of MAG and the BMF Kelso BikeFest (Figure 5.1).

As well as organisations who have the aim of protecting riders' rights there are also others that are more specifically targeted. For example, the National Association for Bikers with a Disability (NABD) is a charity that helps riders who are disabled. Their main aim is not political, but rather practical, offering help for disabled riders to adapt bikes so that they can ride them (NABD 2008a). The NABD also organise rallies and other events such as 'NABD by the Trossachs' held in Callendar, Central Scotland (NABD 2008b).

Figure 5.1 Kelso Bikefest 2007

Charity Works

Despite the image that bikers are rough, tough and anti-social, many take part in charity events each year. For example, at Easter time each year there are a plethora of 'Easter Egg Runs' where riders meet up after a ride to donate Easter Eggs to deserving children. The 2007 'Annual Motorcycle Action Group Easter Egg Run' attracted around 7,500 riders and was described by the BBC as:

> Thousands of 'biker bunnies' descended on Glasgow to deliver an Easter treat for sick children. (BBC 2007e)

One other way that riders may also contribute to society is by using their riding skills for good. For example, there are organisations that do transportation runs for hospitals, such as SERV and the Freewheelers based in Yeovil, Somerset. These organisations aim to move blood, organs, emergency medical equipment and similar products, quickly and safely between medical establishments. These charities do this with minimal, if any, expense to the health service, thus releasing funds for patient care (BBC 2008a).

The above is only a brief discussion of some of the charity activities that motorcyclists are involved in. As these activities tend to only receive minor mention in the media, there is little awareness of these activities, therefore these deeds do little, if anything, to overturn the publicly held view of 'bikers'.

Motorcyclists in the Media

Motorcyclists are portrayed in many ways within the media and in many different lights.

The Radio Four programme *The Archers* has had a character, Alan Franks, the local vicar, who also rides a motorcycle. This is an example of how biking is used in an unusual manner for entertainment value, but at least in this case the image projected is positive (BBC 2005).

A less positive image of riders comes from an unlikely media source, *The Simpsons*.

In series 11 episode 8, entitled 'Take My Wife, Sleaze', Homer wins a bike and forms a motorcycle gang called 'The Hells Satans' (Hells Satans 2008). This episode feeds on the stereotypical image of riders, for example, Homer says:

> Yeah, that's the life for me, Marge. Cruising and hassling shopkeepers.

The outlaw image is further reinforced within this episode when another group of bikers, also called the Hells Satans, take offence at Homer using the name. This gang act in a stereotypical gang manner, wrecking Homer's house and kidnapping his wife. The fact that an iconic cartoon series can use such stereotypical images of riders to good effect demonstrates how much the negative rider image is ingrained within society.

The media coverage of rider-related events certainly help to maintain some of the negative aspects of 'biker' image. Three news items give an example of this. One headline was 'Shot fired by biker robbers' (*Lancashire Evening Telegraph* 2004), another 'Armed biker takes church donation' (BBC 2007c) and the third headline for a story concerning a robber who used a moped as a getaway vehicle was 'Armed biker in raid on jewellers' (BBC 2007b). All of these news items mention that someone who was on a form of PTW was involved in the committed crime and in each case the headline refers to them as 'bikers' (McDonald-Walker 2000). One question that can be asked is would the fact that a car was involved in a robbery be mentioned in the headline?

However, within the entertainment media some positive images of riding have emerged. *The Hairy Bikers* BBC show (BBC 2008e) is one such example, attracting an audience of 3.4 million viewers in January 2008 (*The Hairy Bikers* 2008). This television programme shows two bikers, David Myers and Simon King, riding around various parts of the world and cooking food, often of a local variety, wherever they stop. Motorcycling was also shown in a positive light by the two televised trips, with accompanying books and DVDs, undertaken by Ewan McGregor and Charley Boorman. These two programmes, *The Long Way Round* and *The Long Way Down*, were well received and increased sales of the BMW bikes that were used for these trips. The *Daily Telegraph* (2007) stated that:

> The reason for the big German twin's phenomenal success in the Japanese sports bike-obsessed British market is usually put down to *The Long Way Round*, the televised trip from London to New York two years ago – via Europe and Asia – by Ewan McGregor and friend Charlie Boorman. The pair made the arduous journey on two R1150GS Adventures, giving the bikes massive credibility as well as coverage for their very real off-road ability.

The entertainment media is now starting to show biking and riders in a more positive light. If the general public start to make positive connections with 'biking' from the images being shown in the media then it can be expected that bikes, and biking, will start to appear more in marketing and advertising. The amount of use of positive biking images in marketing can be used as a litmus test with respect to how much public attitude has changed.

The links between celebrity and motorbiking has been a long-standing one with many celebrities consciously using bikes to give them an edgy image. Although there are many celebrities who are attracted to motorbikes in a similar way to other riders. There are some for whom the image is the only attraction. A recent article in the online magazine Luxist suggests that despite Paris Hilton investing $250,000 on a custom motorbike, she has never been seen off the red carpet (Woollard 2008). This is likely to have two results: the lure of the 'celeb lifestyle' may encourage people who would not otherwise ride to become riders; but if it is seen as an anarchistic statement, then it may lead to inappropriate use of PTWs.

Riders and their Attitudes toTraffic Law

In New York, the Police Department's implementation of Mayor Giuliani's 'Zero Tolerance' policy successfully reduced subway crime by catching 'turnstile jumpers'. It was argued that a person who planned to commit a crime on the subway, such as rob a passenger, would not buy a ticket (Cunneen 1999). If riders are the 'outlaws' they are portrayed as, then it would be expected that they would not comply with traffic law.

Vehicle (Excise) Tax

The British Government's review of excise duty evasion (Nasctmento Silva 2007) estimated that around 38 per cent of all motorcycles being used on the public highway were being used illegally due to not being taxed. The estimated tax evasion for all vehicles was 6.2 per cent. These figures also estimated that 32 per cent of all untaxed vehicles on the road were motorcycles, costing £28 million pounds in lost revenue. This figure seemed to justify the notion that riders do show outlaw tendencies, and reflecting this Edward Leigh, the chairman of the Commons Public Accounts Committee said:

> Motorcyclists are particularly liable to evade road tax. Nearly 40 per cent of motorcycles are now unlicensed. If the DVLA's motorcycle enforcement regime is not to be a complete laughing stock, then the Agency and the Department must make the most of new powers to enforce VED off public roads – and strongly consider more severe measures such as impounding unlicensed motorcycles. Large parts of the biking community are cocking a snook at the law. (Committee of Public Accounts 2008)

Paul Routledge of the *Daily Mirror* newspaper commented on the figures stating that:

> Motorcyclists are a pain in the exhaust pipe. They drive recklessly and kill themselves in alarming numbers.... But I didn't know until recently that they also break the law by refusing to pay road tax – apparently 40 per cent of motorcyclists are on the road without a tax disc. (Routledge 2008)

The response from rider and industry groups was to question the figures and the methods used in their calculation. A spokesperson for the Motorcycle Industry Association (MCIA) stated that:

> As a tool for informing policy makers about targeting resources to reduce VED evasion in respect of motorcycles, the current survey appears deeply flawed. (BBC 2008d)

A subsequent review of the research revealed serious flaws in the methodology. The revised calculation estimated the percentage of evaders at 6.5 per cent (DfT 2008; Stone 2008), very similar to that of car evasion (6.2 per cent). This example indicates the willingness of some media representatives to use any story to reinforce the biker stereotype. While this story was covered by a large cross-section of the media, the corrected figures, and subsequent apology, received much less attention.

Other law breaking behaviour that might be expected of 'bikers', such as riding under the influence of alcohol and speeding, are now considered.

Drink Driving/Riding

Drinking may be associated with riding, especially when the activities at rallies and other motorcycle events are considered, however, riders are less likely to fail a breath test than other road users (DfT 2006a). The failure rate for motorcyclists is 1.6 per cent, compared to a 2.0 per cent rate for all road users. However, in other countries such as the United States of America motorcyclists riding while intoxicated is a much bigger problem (Huang and Preston 2004; Hurt, Ouellet and Thom 1981). The evidence suggests that within the UK those who ride understand that operating a motorcycle is complex (Elliott, Baughan, Broughton, Chinn, Grayson, Knowles, Smith and Simpson 2003; Mannering and Grodsky 1995) and impairment significantly adds to the rider's risk. Thus they are less likely than other road users to be over the blood alcohol concentration (BAC) limit (Broughton 2007).

Speeding

Speed is often linked to motorcycling and there are often complaints about PTWs speeding through towns and villages. One person expressed the view:

> Motorbikers are lunatics! ... Motorbikers are immune to speed regulations. (Broughton 2007)

Motorists also sometimes complain that bikes suddenly appear behind them, as expressed by one driver:

> I do my best to look out for them on the road but it is not always possible when they come from nowhere. (Broughton 2007)

The DfT commissioned a project on high unsafe speed accident reduction (HUSSAR) where an understanding of why crashes happen, and to whom, was sought in order to aid in crafting appropriate interventions. This project included the collection of survey data from both drivers and riders. Part of this project also included a focus group comprising riders (Hannigan, Fuller, Bates, Gormley, Stradling, Broughton, Kinnear and O'Dolan 2008) and a national survey of UK drivers, with a smaller survey of motorcyclists.

One major difference that comes from the data is how speed limits within and outwith towns are viewed. Only 4 per cent of riders stated that they would go above the speed limit in town if they thought it was safe to do so, compared to 16 per cent of drivers. When a similar question is asked about exceeding the speed limit when out of town then the position is switched with 66 per cent of riders willing to exceed the limit, and only 40 per cent of drivers. One rider expressed:

> Some bits you go because you know you can go … test yourself and the bike, without necessarily going too fast, but its challenging, there are other bits where you can go where you can just go, bloody fast. (Hannigan, Fuller, Bates, Gormley, Stradling, Broughton, Kinnear and O'Dolan 2008)

However, the figure of 4 per cent of riders who admit to riding faster than the speed limit may have to be viewed with caution, as when asked about their behaviour just before an in-town speed enforcement camera 19 per cent said that they would be doing more than 30 mph, but a similar phenomenon is also seen for drivers. This may reflect that riders, and drivers, may view that some slowing down when entering a speed restricted zone is complying with the speed limit, at least in principle.

These data suggests that riders are more likely than drivers to respect urban limits, but this should not be surprising due to the hazards posed by third-party vehicles within this environment and the vulnerability of bike riders. When out on the open road the level of threat from other vehicles is perceived to be less and therefore riders do not feel so constrained to the speed limits. The figures published by the Department for Transport (DfT 2007a) show that riders speed more on single carriageway roads with a 60 mph speed limit when compared to car drivers. The extreme side of this open road speeding is demonstrated by some press reports, for example, the rider who filmed himself doing 189 mph (BBC 2008c). However, most excess speed on the open roads is at about 10 mph, or so, above the limit.

While riders are not the perfect road users with respect to complying with road traffic law, all the evidence suggests that, contrary to their 'bad boy' image, they do not transgress significantly more than other road users, such as car drivers and in-town probably transgress less.

What do Others Think?

The popular culture image of bikers as 'Wild One' rebels or law breaking Hell's Angels was discussed at the start of this chapter. Does this image still influence now non-riders view those who ride? Another issue that may affect what others think about riders is that PTWs exist in a four-wheel society (Gutkind 2008). In order to ascertain whether such images were prevalent amongst the general population, a survey was carried out asking members of the public what they thought about those who rode PTWs (Broughton 2007). Views on riders were solicited via an open-ended question and the responses were categorised into 23 themes, using a method based upon Miller and Crabtree (1992); some respondents' answers reflected more than one theme. The themes were also categorised into 'negative' (Table 5.1) and 'positive' (Table 5.2) comments. The responses of over two-thirds (71 per cent) of the respondents reflected negative views with the majority commenting that bikes are dangerous or ridden in a manner that makes them dangerous; although comments on enjoyment and the practical elements of riding also featured.

Table 5.1 Negative themes on bikers

Theme	Respondents
Bikes are dangerous	44%
Risk takers/reckless	27%
Do not like bikes weaving/filtering	13%
Riders have a bad attitude/no consideration	12%
Riders have no respect for traffic laws	10%
Bikes are not easily seen	9%
Bikes are noisy	7%
Vulnerable	6%
Riders need to be restricted	2%
Riders are intimidating	2%
Riders are thugs	2%
Riders blame cars for crashes	1%
Bikes are not environmentally friendly	1%
Riding would not be enjoyable	1%

Table 5.2 Positive themes on bikers

Theme	Respondents
Riding is fun	15%
Bikes are practical	13%
Riders have good skills	12%
Riders have good camaraderie	8%
Riders are brave	4%
Other vehicles cause bike crashes	3%
Riders are sensible	3%
Riders are passionate	2%
Riders are OK/good people	1%

Within this survey information was also sought on whether respondents held a motorcycle licence, if they have ever ridden a motorcycle on the public road or if any of their friends or family ride. The data suggest that those who hold a bike licence are more positive towards biking than those who do not with a similar pattern evident for those who have ridden a motorcycle on the public roads in the past. Those who have friends and family that ride also have a more positive view of riders, but this is not as distinct as in the other two categories.

Exploration of these themes suggests that although the extreme image of riders as 'bad boy' renegades may not be held by members of the general public, some still hold the view that riders/biking is a reckless/dangerous activity, with this view being stronger if the respondent does not have personal experience of biking or riders.

Profiling the Rider

According to the DfT (2006a), there are currently around 1.62 million motorcycles within the UK compared to an estimated 33 million cars (DfT 2003a) with bikes accounting for roughly one journey for every 20 car journeys taken. The highest motorcycle ownership rate is in the South West of England and the lowest is in Scotland. In 2004 the ownership rate for Great Britain was lower than any other main EU country (DfT 2006a).

Given the image of bikers it might be expected that riders would tend to have a young, predominantly male, low-income profile. With the exception of being predominantly male, this profile is not generally the case.

Age and Gender of Motorcycle Riders

An age profile for riders is given in the 2006 edition of the Department for Transport compendium of motorcycle statistics (DfT 2006a) showing that riding age is slightly skewed towards the older age groups with the peak age being between 30 and 39 years old. However, over a quarter of riders are aged 50 years or older (Table 5.3).

The DfT compendium of statistics does not present data on the current gender split for riders. However, the 2002 National Traffic Survey stated that males were seven times more likely to make a PTW trip than females (Clarke, Ward, Bartle and Truman 2004), therefore it can be approximated that about 14 per cent of bike riders or passengers are female. Broughton (2007) found a similar gender split (17 per cent) with females being well represented in the younger age groups but the number of female riders reduces at 50 years of age more so than the reduction for male riders. Approximately half of all female riders are aged between 36 and 45; only 7 per cent are over 50, whereas 22 per cent of male riders are aged over 50 (Table 5.4). The lower numbers of females riding at the higher age groupings may be an effect of riding being a physically demanding activity and that older females may feel that they no longer have the strength to control a motorcycle. There may also be a problem of social acceptance for women of this age riding motorcycles, which may mean that this effect will reduce over time.

Table 5.3 Age of riders

Age	% of riders
20 or under	9%
21 to 25	5%
26 to 30	5%
31 to 35	12%
36 to 40	19%
41 to 45	13%
46 to 50	15%
51 to 55	7%
56 to 60	6%
61 or older	8%

Table 5.4 Gender profile by age groups with percentage split of male to female for each age group (χ^2 (9 *df*, n = 1097) = 24.762; p = 0.003)

Age	Male	Female
20 or under	3%	4%
21 to 25	6%	8%
26 to 30	7%	10%
31 to 35	14%	12%
36 to 40	17%	25%
41 to 45	20%	25%
46 to 50	12%	10%
51 to 55	12%	4%
56 to 60	6%	2%
61 or older	4%	0%

Motorcycle Ownership

Broughton (2007) used the motorcycle manufacturer and model information supplied by respondents to classify their bikes into the categories described in Chapter 1. This showed that nearly 70 per cent of respondents rode either sportsbikes, tourers or sports tourers. However, there are differences in the gender split across the bike types with females being over represented in the sports tourer category and underrepresented in tourer bikes (Table 5.5); this may be because tourer bikes are generally heavy machines and females may prefer to opt for a lighter sports tourer. This may also link to the way that females enjoy riding (see Chapter 7).

When bike types were analysed against age there is a progression from the sportier bikes to tourers as the rider gets older (Table 5.6). Younger riders may be attracted to the sports bike due to the glamour associated with high-profile racing events such as British Super Bikes. Especially as, for a relatively modest outlay, a bike can be obtained with a similar performance to that being raced. For example, the Virgin Mobile Yamaha team that competed in the 2007 British Super Bike series rode Yamaha YZF R1 bikes (BSB2006); a road legal version costs under £9,000 (Yamaha Motor Company 2007b). The riding position on a sports bike is hunched over the front of the bike. This position puts pressure on the rider's back and wrists; therefore it is not surprising that as some riders get older they may opt for bikes with a more 'body friendly' riding position.

Table 5.5 Gender by bike type (χ^2 (8 *df*, $n = 998$) = 19.454, $p = 0.013$)

Bike Type	Male	Female	Total
Sports	29%	27%	29%
Sports Tourer	25%	38%	26%
Tourer	24%	13%	22%
Classic/Custom	8%	9%	8%
All Rounder	15%	13%	15%
Total	100%	100%	100%

Table 5.6 Bike type by age (χ^2(8 *df*, n=853) = 80.025, $p < 0.001$)

Bike Type	35 and under	36 to 50	51 and older	Total
Sports	37%	27%	15%	27%
Sports Tourer	25%	29%	25%	27%
Tourer	11%	23%	38%	22%
Classic/Custom	4%	11%	8%	8%
All Rounder	23%	10%	15%	15%
Total	100%	100%	100%	100%

Using the Performance Index (PI) Broughton (2007) found that the percentage of riders who ride high, or very high, performance machines does not vary much with age, however, younger and older riders are more likely to ride low or very low performance machines compared to those in the middle age group (Table 5.7).

The over-representation of younger riders in the very low performance group may be due to the current licensing restrictions that do not allow riders under 21 to ride the more powerful machines (DSA 2004a). Once the age of 21 is reached, it will take time for a young rider to pass the test to gain access to larger machines and then to obtain one. One of the other barriers for young riders wanting a more powerful bike is the cost of insurance with insurance companies generally loading younger rider premiums. Broughton did not find any significant differences when comparing gender with performance index.

Table 5.7 Performance Index against age of respondents (χ^2(1 *df*, *n*=457) = 17.213, *p* = 0.028)

PI	35 and under	36 to 50	51 and older	Total
Very low	24%	11%	16%	16%
Low	14%	22%	29%	21%
Medium	20%	25%	17%	22%
High	21%	21%	21%	21%
Very high	20%	20%	17%	20%
Total	100%	100%	100%	100%

Income and Occupation of PTW Riders

A survey was carried out by the author for this publication where respondents were asked to select a category that indicated their yearly earnings (Table 5.8) and their occupational group (Table 5.9).

The mean earnings of the respondents tended to be towards the upper level of group 5 (mean of 5.79) the £25,000 to £30,000 category, with a median and modal value of 5. If riders who have retired from working are removed from the data then the mean rises to 5.83. Linear interpretation of this group gives an estimated mean earnings figure of £28,950 or £29,150 if those who have retired are excluded from the data.

The national average earnings at the end of 2007 was £457 a week (BBC 2007d). This equates to £23,946 per annum; therefore the survey data suggests that the average rider earns about £5,000 per annum above the national average, around 20 per cent higher.

The social groupings used in this survey utilised categories to match up with the ones used for national statistics, thus allowing a comparison of the rider profile with the national profile (NRS 2006).

Rider earnings are reflected in the occupational profile of the respondents; more than half (53 per cent) of the respondents indicated that they hold middle/upper managerial or a professional position (Table 5.9) against the national figure (NRS 2006) of approximately a quarter. Note that the category of 'lowest levels of subsistence' within the Broughton data includes the unemployed, students, casual workers and those who have retired, as well as some riders categorising themselves as 'other'. Very few riders are from the semi-skilled/unskilled category when compared to the national figure (2 per cent against 15 per cent). However, retired people are over-represented as riders with about one in eight riders fitting into this group.

Table 5.8 Earnings of PTW riders

Group	Earnings	Per cent	Cumulative
1	Under £10K	6%	6%
2	£10K to £15K	14%	20%
3	£15K to £20K	9%	29%
4	£20K to 25K	7%	36%
5	£25K to £30K	16%	52%
6	£30K to £35K	11%	63%
7	£35K to £40K	10%	73%
8	£40K to £45K	5%	79%
9	£45K to £50K	6%	85%
10	£50K to £55K	4%	89%
11	£55K to £60K	1%	90%
12	More than £60K	10%	100%
Total		100%	

Table 5.9 Occupational groupings of respondents

Social Group	Survey Per cent	National Figure
Upper management	21%	4%
Middle management/professional	32%	22%
Junior management/clerical	7%	29%
Skilled manual	16%	21%
Semi-skilled/unskilled	2%	16%
Unemployed	2%	
Student	2%	
Retired	13%	
Other	4%	
Lowest levels of subsistence		8%
Total	100%	100%

The demographics and occupational data suggest a relatively older, affluent, predominately male motorcycle rider in occupations associated with responsibility. This is at odds with the image of motorcyclists often portrayed or held by the media and non rider. Given this relatively affluent profile, it would be interesting to gain an insight into the economic impact of PTW riding and relating activities. How much do motorcyclists spend in the pursuit of their riding activities?

Spending on Motorcycles and Related Activities

Broughton (2007) reported on riders' spending habits. This included not only spending on their bike, but also on consumables (fuel, oil, etc.), accommodation when on biking trips and biking events (for example, rallies such as Kelso Bikefest, bike shows such as the National Bike and Scooter Show held annually at the NEC in Birmingham and racing events like the British Super Bikes). The respondents' average spend per annum on bikes and biking kit was £2,032 with a total of £4,263 per year being spent on bike-related activity. According to the Expenditure and Food Survey (Office for National Statistics 2004), the average household has a weekly spend of £0.60 on PTW purchases and £0.16 on accessories, spares, etc. giving a total spend of £0.76. However, this average spending is calculated for all households, with only 2.3 per cent of households within the UK owning a PTW (DfT 2006a). Therefore the weekly spend for households owning a bike is approximately £33 per week for bike and bike accessory purchases, equating to £1,731 per year. This recalculation of the ONS figures is considerably less than half the spending suggested by Broughton (Table 5.10). This may partly be due to elements of spending associated with PTW use being allocated elsewhere in the Expenditure and Food Survey, for example, spending on accommodation while on biking trips being allocated to leisure/holiday spending in this survey.

Table 5.10 Mean spending on bike-related activities

	Mean	Lower 95% CI	Upper 95% CI	Modal	Median
Spending on bike/bike kit	£2,032	£1,709	£2,355	£1,000	£1,000
Spending on consumables	£1,241	£1,065	£1,418	£1,000	£1,000
Spending on bike-related accommodation	£430	£298	£562	£0	£200
Spending at/on bike events	£216	£162	£270	£0	£100
Other bike related spending	£344	£249	£440	£0	£200
Total spend	£4,263	£3,745	£4,782	£1,500	£3,000

Broughton (2007) reported on the value of respondents' current bike showing that the average value of the respondent's bike was around £3,000 (Table 5.11). In 2004, 2.3 per cent of British households owned a motorcycle compared to 73 per cent of households owning at least one car, with ownership of a bike being more common in households that owned one, or more, cars (DfT 2006a). This indicates that the motorcycle may be a vehicle of choice rather than the main mode of transport for many people. Yet there is a high level of capital commitment on the bike and the ongoing expenditure associated with the machine, despite often being a non-essential/additional transport mode and owned for pleasure rather than necessity.

Spending on Bike-Related Activities

Broughton (2007) asked respondents what they spent on motorcycles, biking equipment and other related activities. The results are shown in Table 5.12 with the average biker spending over £4,000 a year on bikes and bike-related activities, with the median average spend being £3,000. Of this amount around 20 per cent is spent on biking events and accommodation.

When spending on a motorcycle and its related consumables is separated out from monies spent on accommodation and events, it can be seen that the average rider spends £646 per year on biking-related events and accommodation (Table 5.13). However, the modal spend value is zero, indicating that a reasonable number of riders (19 per cent) do not spend money on these types of activities.

Table 5.11 Value of bikes

Bike Value	%	Cumulative
Less than £1,000	8.4%	8.4%
£1,000 to £1,999	19.6%	28.0%
£2,000 to £2,999	19.6%	47.7%
£3,000 to £3,999	16.3%	63.9%
£4,000 to £4,999	13.3%	77.2%
£5,000 to £5,999	6.5%	83.7%
£6,000 to £6,999	5.6%	89.3%
£7,000 to £7,999	3.9%	93.3%
£8,000 to £8,999	2.1%	95.3%
£9,000 to £9,999	0.7%	96.1%
More than £10,000	3.9%	100.0%

Table 5.12 Motorcycle-related spending

	Mean	Modal
Spending on bike/bike kit	£2,032	£1,000
Spending on consumables	£1,241	£1,000
Spending on bike-related accommodation	£430	£0
Spending at/on bike events	£216	£0
Other bike-related spending	£344	£0
Total spend	£4,263	£1,500

Table 5.13 Motorcycle spending on bikes and activities

	Mean	Modal
Spending on bike and running the bike	£3,273	£2,000
Spending on events and accommodation	£646	£0

If those who do not spend any money on motorcycle events and accommodation are removed from the figures then the mean spend rises to £798 per annum (95 per cent CI £617 – £980; median £500; modal £500).

Motorcycle Tourism

For many, motorcycling is an expressive activity (Broughton 2005) and riding is mainly carried out for pleasure. This leisure riding is described by Business Week (1965) as a 'second life', that is a break from normal working life. Some take this 'second life' to the next level and extend their leisure riding to include the activity of motorcycle tourism. One of the major differences in motivation for those on riding holidays to most holiday-makers is the importance of the journey over the destination with good scenery and good roads being important journey features. Motorcycle tourism can take many forms in how it is organised, from individuals or groups of riders arranging their own vacation, to club run tours and those organised by professional companies. The numbers of those doing this type of riding are significant enough for some tourism agencies to take an active interest in this market segment; it is estimated that by attracting just 0.1 per cent of potential UK riders would bring in an additional tourism spend of nearly £5 million pounds (Keys 2006). Agencies responsible for marketing destinations such as Visit Scotland have recognised the potential of this market. They have sought to make riding in Scotland a more pleasurable experience by launching initiatives such as 'Bikers Welcome' (VisitScotland 2006). This initiative actively promotes accommodation that makes special provision for PTW riders, such as hard standing for their bikes and a place to dry protective gear.

The mean spend per year just on accommodation is £430 per person, however, for a business this may be multiplied because those embarking on motorcycle tourism often do so in groups. For the lower age groups (under 40s) the spending on accommodation is higher, with a mean spend of £773 per annum. The accommodation value for the under 40s can be estimated at around £470 million each year as 47 per cent of the 1.62 million riders are under 40 (DfT 2007c) and of these 80 per cent engage in motorcycle tourism, a total of 610,000 riders. Also, Scotland is popular for PTW riders from Europe particularly Holland and Germany due to easy access from ports such as Newcastle and Rosyth. As overseas visitors tend to spend substantially more than domestic tourists (Walker 2007), this has the potential to be a very lucrative market.

VisitScotland has developed initiatives specifically designed to attract and cater for this lucrative market. The 'Bikers Welcome' scheme seeks to encourage accommodation providers to make provision that caters for PTW riders, for example, VisitScotland (2006) states that:

> An area should be available for drying outdoor clothing and footwear at an ambient temperature so clothes can dry overnight.
>
> Wash down facilities for motorbikes should be available for guests use.

Organised motorcycle tours come in two basic formats, self guided and fully guided, for example, White Rose Tours offer a fully guided tour of the Scottish Borders that includes the booking of all accommodation and ferries and access to experienced tour guides (White Rose Tours 2008). Self-guided tours are tours where the rider is in control of the pace and route of the ride with the tour provider organising the accommodation and providing suggestions of what to see. On some tours they will also provide a back-up vehicle in case of bike breakdown and that might also carry your luggage to the next stop as well. One provider described the type of rider who may wish to avail themselves of a self-guided tour (ScotlandByBike 2008):

> Our self-guided tours are designed for our customers who like to travel on their own or with friends in a small group. If you are an experienced biker and traveller who doesn't feel the need to be guided, then you'll find the self-guided tours suitable for you.

For trips abroad, fly-ride can be offered with the rider being able to pick up a hire bike at the country of destination. Some firms even make a feature of the hire bike, for example, West Coast Harley Tours offer the opportunity to tour the Canadian Rocky Mountains on a Harley Davidson (West Coast Harely Tours 2008).

Another aspect of motorcycle tourism is motorcycle events such as custom bike shows and rallies.

Events and Rallies

The 'Rides and Rallies' website (http://www.ridesandrallies.co.uk) lists over a hundred rallies in the United Kingdom for 2008. Some riders faithfully attend their favourite event every year, some attend events on an ad hoc basis and others attend as many events as possible with virtually every weekend over the summer period revolving around a rally.

These rallies can vary in size from minor events attracting less than a hundred through to major events attracting many thousand, for example, the North West 200 in Northern Ireland attracts an estimated 100,000 visitors many of whom are riders. They often have entertainment in the form of live music and/or a disco, camping facilities for those staying overnight and, of course, a bar (see Figure 5.1).

A UK Bike Forum member stated that she went to rallies because of the:

> Good music, good food (because we cook our own) and great company. There's nothing better than meeting up with like minded individuals and just having a laugh. It's rare that I come away from a rally without having met a few new rally mates. It also means that I maybe get to go to parts of the UK that I've not visited before.

This again illustrates the sense of 'community' as discussed earlier while also indicating the secondary nature of the actual destination.

Who Spends What?

PTW riders are not a homogenous group and vary considerably in profile. Analyses of data gathered by Broughton (2007) suggest that spending can be linked to demographics.

Younger riders (under 40) spend considerably more than older riders, with older bikers spending a mean value of £526.79 compared to £805.00 for the under 40s ($t(243) = 2.22$, $p = 0.027$). There was no significant difference between total spending and spending upon the bike when compared with age.

If those who do not spend any money on events and related accommodation are excluded from the analysis then higher levels of spending for the under 40s group is seen. The mean spend for the under 40s on accommodation is £773 per annum compared to £357 for the over 40s ($t(197) = 2.522$; $p = 0.012$). This higher level of spend is also seen in the combined events/accommodation spend with the mean under 40s spending £1,032 compared to £628 for the over 40s ($t(197) = 2.160$; $p = 0.032$).

Not surprisingly, those who earn under £30,000 spend less on their bike and bike kit compared to those earning more than £30,000 (£1,576 compared to £2,631 – $t(242) = -3.20$; $p = 0.002$). Those in the lower, under £30,000, earnings bracket also tend to spend less on consumables (£1,094 compared to £1,443 – $t(242) = -1.91$; $p = 0.057$) as well as spending less overall (£3,496 compared to £5,273 – ($t(242) = -3.37$; $p = 0.001$).

Those who earn over £30,000 a year are more likely to spend money on motorcycle events and accommodation, while a quarter of those under £30,000 per annum will not spend any money in this area (Table 5.14).

When the average spend is calculated with those who do not spend any money at motorcycle events, or on related accommodation, then the mean events-related spending for the higher earners is £356 per annum compared to £179 ($t(195) = -2.665$; $p = 0.008$).

A similar pattern to earnings is seen when social groupings are analysed with those in upper/middle management or professional occupations spending more in total, £4,706 compared to £3,512 ($t(243) = 2.05$; $p = 0.041$). This higher social group also spends more on their motorcycle and its associated costs than those in the lower social groupings, £3,529 compared to £2,786 ($t(243) = 2.05$; $p = 0.040$).

Spending and 'Biker Image'

The image of motorcycle riders is often one of outcasts from society; however, riders are more likely to be in the higher earning brackets with a mean level of earnings around £5,000 pa higher than the national average. They are also likely to have a job encompassing a high level of responsibility with over half of riders holding a middle/upper managerial or a professional position compared to a national figure of a quarter.

Along with the high earnings also goes a large disposable income and for around 80 per cent of the riders some of this is spent on motorcycle tourism, with this activity being more prevalent for those who earn over £30,000 per annum.

Table 5.14 Events spending for those earning over £30,000 pa (χ^2 (*1* df, $n = 244$) = 7.357, $p = 0.007$)

	Do not spend	Spend	Total
Earning under £30K	26%	74%	100%
Earning over £30K	12%	88%	100%
Total	19%	81%	100%

Despite the image that would seem to be prevalent amongst the general population that 'riders' are dangerous and irresponsible, the profile of the PTW riders from this research suggest that they are likely to be middle-aged, be in positions of responsibility in the workplace and be relatively affluent. This profile, together with PTWs not being the sole transportation in most households, suggests that 'biking' has become a middle class hobby rather than a cheap/alternative mode of transport for those on lower incomes. Nor is it the 'young rebels' who are riding the most powerful bikes, but those in the middle age ranges. If the image of the 'outlaw' risk taking young rider evading social norms and seeking danger is not the case, then what are the motivations for riding PTWs? The following chapter explores reasons and motivations for being a PTW user.

Chapter 6
Why do People Choose to Ride?

The discussion so far has been centred on issues such as who rides and the social environment within which they ride. Given the relative danger of the motorcycle as a mode of transport (DfT 2007c; RoSPA 2001), there has to be some discussion on the motivations for riding. Having dispelled some of the myths associated with the image of bikers and their 'outlaw' reputation, the reasons for riding will be explored in this chapter.

Earlier discussions have explored the relatively high skill base required for riding a PTW compared to driving a car. This combined with increased vulnerability would suggest that there is a challenge in riding not as readily found in driving. The traditional image of bikers is as reckless thrill seekers who actively seek danger and live 'on the edge'. These views were expressed by some non-riders who said (Broughton 2007: pp. 410–413):

> …they are risk takers.
>
> I just think that it is a very dangerous activity and that motorbikers are taking very big risks.
>
> I get this mental image of these big, white guys with leather jackets with Hells Angels signs on their back and great big colourful tattoos who ride Harley Davidson's.

In the chapters on theories of Risk Homeostasis (Chapters 3 and 4) there was some discussion on potential reasons for riding and why risk may not be the reason for riding. This chapter explores the reasons and motivations for riding from the perspective of the Powered Two-Wheeler (PTW) user themselves.

Riding Trip Purpose

The National Traffic Survey (DfT 2003a) categorises 'reasons for trips' into various purposes including: work/business/education; shopping; visiting friends; and other leisure (Table 6.1).

Although the 'work/business/education' category showed similarities between car and bike use, in other leisure trips bikes were used less frequently but the trips were almost double the length. This implies that the actual ride was the attractor, not the destination.

Table 6.1 Trip reasons for car and PTW vehicles, categories from the Department for Transport 2003a

	Motorcycle		Car	
Trip reason	**Trips per rider per week**	**Average trip length (miles)**	**Trips per driver per week**	**Average trip length (miles)**
Other leisure	0.7	24.9	1.6	12.8
Visit friends	0.9	10.6	2.3	10.6
Work/business/education	5.1	9.5	4.7	11.5
Shopping	0.7	4.3	3.3	5.2
All trips	8.0	10.5	16.4	8.4

The figures from the Department for Transport (DfT) indicate the purpose of the trip, but the data does not explore the reason why the motorcycle was used. This is an important distinction, as for most riders there is a car within the household (DfT 2006a). This suggests that the bike is probably being used as a vehicle of choice rather than necessity for most trips. So why do they choose a PTW over other available modes of transport?

Reasons for Riding

Two of the main purposes for riding can be categorised as function-related, such as commuting, or leisure-related (Broughton and Stradling 2005).

Commuting

The National Traffic Survey reports that around two-thirds of respondents commute to work by motorcycle (DfT 2003a) and, of those who use their bike to get to work, only a small percentage (6 per cent) do so because they feel that they have no other option. Thus, the majority of those who use their bike to get to work do so because they enjoy the ride (Broughton 2007). One commuting rider said:

> It is great fun, even in the wet. (Broughton 2007: p. 292)

Broughton (2007) reported that about a third of commuting riders did so mainly for the convenience. One of these riders expressed:

> Being able to get there at my speed. No frustration of the cooped up car driver in a queue. (Broughton 2007: p. 291)

Other reasons why some used their bike to commute to their place of employment are the economics of running a machine; easier, often cheaper, parking; and the ease of access through traffic resulting in reduced journey times (City of York Council 2005; National Motorcycle Council 2000).

These factors are often seen as important reasons for PTW use. Some UK local authorities recognise these factors and can see benefits in encouraging PTW use in their area, for example, the Staffordshire County Council's (2005) local transport plan states that:

> Powered Two-Wheelers (PTW's) offer the same potential for personal mobility as private cars whilst contributing less congestion, pollution and damage to roads. PTW's are not subject to the same delays in congested traffic and so spend less time wasting fuel idling in queues. They are lighter, generally more fuel-efficient and take up less space, whether parked or moving. (Staffordshire County Council 2005)

Leisure Riding

Cornwall County Council conducted a survey amongst PTW riders in their area. Within this questionnaire one question asked about trip purpose; only 1 per cent stated that they never used their bike 'purely for fun', with 49 per cent riding for fun at least two to three times a week and 15 per cent riding purely for fun everyday (Cornwall County Council 2004). From the data presented by Broughton (2007) and Cornwall County Council (2004) it can be concluded that fun and enjoyment are major reasons for riding.

The results from analysis of data collected by Broughton (2005) and examined with the data from the Cornish survey (Cornwall County Council 2004) establishes that riding is predominantly a leisure activity. The expressive nature of riding is a major difference to the predominately functional driving of cars. Most riders do some leisure riding, with Broughton (2007) reporting that only 4 per cent of riders claim not to carry out leisure riding; this 4 per cent consisted nearly exclusively of males.

The importance of leisure riding is illustrated by leisure riding trips being more that twice the distance as the average trip length (DfT 2003a). Enjoyment gained while riding is an important factor in bike use. What though is enjoyment and how might this relate to riding?

Theories of Enjoyment

The verb to enjoy is defined by the *Oxford Concise Dictionary* (2001) as:

> To take pleasure in. (*Oxford Concise Dictionary* 2001)

Pleasure, taken from the verb 'to please', is described as:

> A feeling of happy satisfaction, the state or feeling of being pleased or gratified. (*Oxford Concise Dictionary* 2001)

Happiness is also related to enjoyment and pleasure:

> Feeling or showing pleasure or contentment. (*Oxford Concise Dictionary* 2001)

What though can cause enjoyment or make one happy? Aristotle taught that living is best regarded as 'a longing and desire for a good life' and that people want to do good things, live well and to do well: that is people have a desire to live a happy and enjoyable life. He also said that:

> Everything we do is ultimately aimed at experiencing happiness.

We pursue other things, like wealth, love or fame in the hope that these will make us happy. Happiness is different from these things as we do not seek happiness to gain anything else; happiness is the end state or the goal that we seek. The aim of being happy is even enshrined in the American Declaration of Independence written in 1776:

> We hold these truths to be self-evident, that all men are created equal, that they are endowed, by their Creator, with certain unalienable Rights, that among these are Life, Liberty, and the pursuit of Happiness.

Therefore it can be argued that man has the right to pursue his happiness, pleasure and enjoyment; for some this pursuit may come in the form of riding.

Lyubomirsky, Schkade and Sheldon (2005) suggest that happiness is controlled by three main factors: a genetically determined set-point of happiness; happiness relevant circumstantial factors; and activity-related practices, with the activity factors offering the best possibilities for a sustained increase in a person's happiness.

The set-point model, sometimes called the 'hedonic treadmill' (Brickman and Campbell 1971), gives the idea that every person has a set point of happiness that they will return to after an event that either lowers (such as a death of a loved one) or raises (such as getting married) their happiness level (Csikszentmihalyi and Hunter 2003). Kammann (1983) expressed this idea as:

> Objective life circumstances have a negligible role to play in a theory of happiness. (Kammann 1983)

Riding may be more to do with subjective rather than objective circumstances and therefore a method that riders employ to raise their happiness level back to, or above, their set point level. Even the anticipation of a ride can raise a person's enjoyment level, and in time this can produce a higher happiness set point (Segerstorm 2006).

Riding can be enjoyed by a large selection of the population, for example, one rider said that motorcycling was:

> Cheap accessible performance.

Enjoyment and Resources

Financial wealth is often associated with happiness and enjoyment of life. The economic view is that well-being and happiness depend on 'life's circumstances'; for some this means that happiness is directly related to gross domestic product (GDP) per capita (Easterlin 1995). Recent research indicates that this is not always the case and in some countries the trend in well-being and happiness has not increased with GDP, rather it has remained constant despite GDP increasing (Easterlin 2005). This suggests that subjective well-being (SWB) increases as affluence rises until a certain level is achieved, then SWB no longer rises with income (Walker 2007). However, the use of resources can bring happiness. Van Boven (2005) discusses the use of resources when used to gain life experiences can have an impact on SWB:

> Allocating discretionary resources toward life experiences makes people happier than allocating discretionary resources toward material possessions. (Van Boven 2005)

Many riders will say that riding certainly gives them life experiences. The use of resources for enjoyment can be illustrated by the pleasure that some people get from shopping where enjoyment or pleasure can be obtained from emotional satisfaction when a shopper hunts for, and obtains a bargain (Schindler 1989) as well as giving a feeling of pride, intelligence and a sense of achievement (Mano and Elliott 1997). The enjoyment gained from finding a bargain may also be caused by 'beating the system' (Morris 1987). It is also sometimes argued that enjoyment from the shopping experience comes from the change of environment, that is, 'getting out of the house' (Lehoten and Maenpaa 1997).

These enjoyment models can be applied to riding. A rider must employ resources to take part in the activity of riding, with the financial resources including the bike and fuel as well as expending monies on excise tax and insurance. It is not only financial resources that are employed as the rider will also have to expend energy and effort while riding. This will be at a higher level than when driving as riding is more physical and demands a higher skill base. A rider will also spend time resources on the activity to enable enjoyment. Enjoyment for some may also come from the feeling of beating the system and a sense of achievement, for example, by manoeuvring their bike past a long stream of stationary traffic often referred to as filtering. This was expressed by riders who stated that they found enjoyment via:

> Getting through traffic queues (filtering) more quickly than in a car.
>
> Access through traffic.
>
> Being able to get there at my speed. No frustration of the cooped up car driver in a queue.

The Riding Environment

Riding can also involve moving into a different environment, a recognised way of increasing enjoyment (Lehoten and Maenpaa 1997). The environment itself can give enjoyment: Three riders expressed it as:

> You can smell the countryside.
>
> Freedom of not being enclosed, enjoyment of being able to experience the countryside … first hand. To be able to smell the fields you travel beside.
>
> Just to get out on the bike and go to places that you otherwise would not get to.

Social interaction and friendship can be a source of enjoyment such as spending time on joint leisure activities (Argyle and Hills 2000). Many activities are made enjoyable or more enjoyable due to the opportunities for social interaction; it has even been suggested that some of the enjoyment that can be found in shopping may be due to people seeking out social interaction, Tauber (1972) for example, found this to be a key factor in the shopping experience. There can also be a social enjoyment element for those who take part in activities that are mainly solitary, for example, gardeners and collectors who may get their social pleasure from occasional meetings or reading club magazines (Hills, Argyle and Reeves 2000). Riding is in some ways a solitary activity with a rider being isolated from others; however, many have noted that there is a social nature to riding as well. One rider summed up the enjoyable aspect of the solitude stating that enjoyment was found by:

> Being alone in my head with no one else talking to me.

Yet many riders also talked of the feeling of comradeship within the biking community, about being able to share stories with other riders and the friendliness that exists between riders: One rider said that she gained enjoyment from:

> The friendship of other bikers.

One other rider expressed:

> Most bikers, including myself, always give a wave in passing.

It is not being suggested that enjoyment cannot be found in the driving activity, rather that the goal of driving is mainly functional with a level of enjoyment while riding is expressive with a level of functionality. The Royal Society for the Prevention of Accidents (RoSPA) (2008) acknowledges that driving can be enjoyable:

> Improved driving skills will not only make you safer, but you'll also find you get a lot more enjoyment out of your car. (RoSPA 2008)

Marketing the Enjoyment of Road Use

This enjoyment of road use is one that has not been lost of those who market cars; Mazda (2007) state that their Mazda3:

> Combines sporty driving enjoyment with low fuel consumption.

The literature that Ford (2007) distributes on the Fiesta states that:

> Throughout each generation, the Fiesta badge has always been synonymous with agile, entertaining handling and this remains the case. The bold new design may be tougher and safer but this isn't at the sacrifice of driving enjoyment.

The RAC Foundation (2002) in discussing the future of driving also acknowledges that enjoyment for some is an important element of driving:

> It is also likely that a small separate 'class' of vehicle, designed to cater for those who drive for enjoyment rather than transportation, may have been developed using more traditional engineering.

A similar picture is seen within motorcycling, with being better trained seen as allowing the rider to get more enjoyment out of riding (BikeSafe 2008):

> BikeSafe is an initiative run by Police Forces around the United Kingdom who work with the whole of the biking world to help to lower the number of motorcycle rider casualties. By passing on their knowledge, skills and experience, police motorcyclists can help you become a safer more competent rider. They help you to increase your ability and confidence, so you can get even more enjoyment from riding your motorcycle.

Enjoyment also features in the marketing of motorbikes, in a similar way to cars. The Yamaha Motor Company (2007a), in an article on their MT 03 machine states that:

> Yamaha's goal is to bring excitement and enjoyment to its customers. Today, many people don't have the time to go to a beautiful area to ride their bike, they want to enjoy it immediately out of their garage – even if there is traffic, even if the roads are not so great. So Yamaha wants to offer a lot of elemental motorcycling fun in a short ride.

Enjoyment seems to be an important element of road use; if it was not then marketers would not be expending resources to give the public the expectation of enjoyment and fun as an aid to selling their products. The enjoyment they are trying to sell is a long lasting one that begins before the start of the journey and remains to the end and beyond.

Enjoyment is not 'just an instant in time' rather it is related to a period of time (Griffin 2002) and an activity that makes one happy would occur over a period of time. Therefore it is not surprising that enjoyment is also often related to participation in sport. Scanlan and Simons (1992) define sports enjoyment as a:

> Positive affective response to the sport experience that reflects generalized feelings such as pleasure, liking, and fun. (Scanlan and Simons 1992: pp. 203–204)

Riding as a Sport?

Some of the pleasure that can be achieved via sport is to do with the intrinsic motivation that is obtained from competence and self-determination (Deci and Ryan 1985). Wankel and Kreisel (1985) comment that extrinsic factors such as winning were also important for gaining enjoyment from sport. Within the sports research literature, movement sensations (Scanlan, Stein and Ravizza 1991) and competence have also been identified as sources of enjoyment (Scanlan and Lewthwaite 1984; Wankel and Kreisel 1985).

Riding has a lot of sports-related elements, such as the need for competence, with the competence and skill level being higher than those required for driving cars. The rider has self-determination; how a particular trip goes is largely down to the choices that the rider makes such as the chosen route and the manner it is ridden.

Movement sensations are linked very closely with riding, with the bike having to lean when negotiating corners. The kinaesthetic sensations; the physical feelings felt and controlled by the rider also add to the feeling of pleasure that can be gained from riding:

> The freedom and the fresh air, and of course cornering has to come into it.

The winning element as a need for enjoyment is an interesting facet, especially when many riders do ride as a group and some riders may feel the need to compete with their peers.

Examining riding against other leisure activities suggests many similarities in how PTW users may gain their enjoyment through aspects of riding such as the sensory, physical and social satisfactions gained.

'Likes' and 'Dislikes' of Biking

In a survey of Scottish riders by Broughton (2007) respondents were asked what their 'likes' and 'dislikes' were about riding. The responses were categorised into 15 'likes' and 10 'dislikes' using a method based upon Miller and Crabtree (1992). Comments reflecting the 'likes of riding' made up 57 per cent of the responses.

Elements Riders Liked About Riding

Table 6.2 lists the most common reasons why riders like to ride with 'Freedom' being the most common reason for riding. The sense of belonging to 'the biking community' was also considered important as illustrated in these quotes taken from the survey:

> Being part and feeling part of the biking community, all the biking events, races, rallies, runs, etc. Mutual respect between bikers.

> All bikers I have met are so nice, the fact that most bikers, including myself, always give a wave in passing.

The convenience of riding a bike is one of the major 'likes' for riders with convenience taking several forms such as access through traffic and ease of parking, as seen from these quotes from the survey:

> Getting through traffic queues (filtering) more quickly than in a car.

> Ease of parking and the ability to avoid hold-ups. (Broughton 2007)

Table 6.2 Rider 'likes'

Likes	%
Freedom	22%
Camaraderie/Social	16%
Convenience	14%
Excitement	8%
Fresh air/Nature/Scenery/Places	7%
Speed	6%
Enjoyment	4%
Mechanics	4%
Solitude	4%
Use of skills	4%
Economics	3%
Drivers/People	1%
Other	8%
Total	100%

Some of the respondents commented that one of their 'likes' was using their riding skills (4 per cent); for example:

> The kick from co-ordination in using a m/cycle – balance, speed, judgement.

> Satisfaction of control and use of skill.

Elements that may relate to risk, such as speed (6 per cent) and excitement (8 per cent) did not appear often, rather the 'likes' of riding are mainly related to gaining enjoyment, such as freedom and social elements, as well as the convenience of using a motorcycle.

Enjoyment of the surroundings also featured in respondents' comments. As riders are not encased in a box like car drivers they are in a better position to experience the fresh air and scenery. One rider expressed:

> The roads and scenery in Scotland, it's a great way to explore and you gain total appreciation of the country (and I'm English).

Rider 'Dislikes'

Within the research carried out by Broughton (Broughton 2005; Broughton 2007; Broughton and Stradling 2005) riders were also asked to give their 'dislikes' (Table 6.3) of riding. One of the 'likes' about riding was to get out in the fresh air and enjoy nature, but this 'like' has a flip side in bad weather, and by far the most 'disliked' thing concerning biking was bad weather. Nearly half of respondents commented on this, for example:

> Being wet in summer.

> Cold wet and bloody miserable winters.

Another area of dislike often expressed was against another road user. The majority of riders did not give a blanket disliking, rather their distain was reserved for those who put themselves, or other road users, in danger, although one respondent did single out 'Volvo drivers' for particular attention. Riders, in general, seem to dislike those who drive cars more than other road users. As the main mode of transport on public roads is the car, it is more likely that riders will have had a near miss, or another negative experience, involving a car and this may have prejudiced their judgement. Some comments that were aimed specifically at car drivers were:

Table 6.3 Riding general 'dislikes'

Dislikes	%
Weather	23%
Car drivers	20%
Poor/bad road surface	9%
Law enforcement	9%
Other road users	9%
Cost	8%
Others attitude to riders	6%
Poor bike/kit quality	3%
Congestion	2%
Other	9%
Total	100%

Source: Broughton 2007.

> Lack of space/distance by some car drivers.
>
> Bad car drivers take advantage of the fact that you are more vulnerable than them.
>
> Careless inobservant drivers.

Where comments that were more generally about road users included:

> Other road users who are very ignorant and sole purpose seems to be to cause accidents.
>
> Other road users/abusers.
>
> Lack of consideration from other road users.

Motorcyclists prefer a consistent road surface because they have limited tyre contact area on the road making them unstable compared to cars and other vehicles with more wheels (Institute of Highway Incorporated Engineers 2005). This is reflected by the number of riders stating that poor road surface quality is one of their 'dislikes'.

Almost one in 10 commented on law enforcement as a factor that they did not like; speed cameras, attitudes and inconsistencies of the police and police forces were included in this theme. These particular dislikes are expressed by one respondent who said:

> Difference of police forces attitude that is, one booking for a noisy can or small number plate, and another saying that noisy cans and small number plates didn't kill anyone.

A very small proportion of the riders surveyed complained that they found congestion a problem, and then it was often a specific congestion problem as expressed by one rider:

> The roads can get choked with tourists, caravans and sheep.

The 'other' category was used for the more unusual responses that could not be placed in a theme, such as 'Sheep' and 'Insects'.

The part of the survey seeking rider 'likes' found that convenience including access through traffic was a major plus, so the fact that only a small number state that congestion is a problem should not be surprising. It is interesting to note that the responses for 'likes' were considerably more than for 'dislikes'.

As motorcycle use is more dangerous than most other forms of transport, it is often thought that riders enjoy risk and that this attracts 'thrill seekers' (Mannering and Grodsky 1995). How was risk categorised within the 'likes' and 'dislikes'?

Risk

It is noticeable that in the responses on 'likes' and 'dislikes', not one respondent mentioned risk as a 'like'. Within the 'dislikes', although risk was not directly mentioned, there were statements regarding vulnerability of riders. Comments indicate that riders are aware that they are at risk and the factors they feel may generate that risk.

Riding for Enjoyment

When asked what they liked about riding, most riders gave answers citing ideas associated with pleasure, such as freedom, or convenience; however, some authors argue that riding a PTW cannot be enjoyable due to the high level of risk involved, considering it 'an extremely risky venture' (Bellaby and Lawrenson 2001). There is also a pervasive public perception that enjoyment is sought, and found, in the high levels of risk that riders face (Broughton 2007).

Most current safety initiatives are founded on the assumption that the goal of the road user is simply to reach their destination safely so that they may then fulfil their trip purpose – work, shop, enjoy a social occasion, etc. (Beck, Wets, Torfs, Mensink, Broek and Janssens 2006). Transport is a method that joins up places where people go so that they can meet their obligations (Stradling 2003); however, a transport mode may also serve affective as well as instrumental functions (Steg 2004; Steg, Vlek and Slotegraaf 2001; Stradling, Meadows and Beatty 2001). The driving of a car, or riding a motorcycle, is a skill-based, rule-governed expressive activity involving ongoing, real-time negotiation with co-present, transient others in order to avoid intersecting trajectories. However, beyond this, the use of a motorcycle may be described as having an expressive function. Many recreational riders will go out on their bike 'just for a run', often without a specific destination in mind except to eventually arrive back home. For this A to A rather than A to B riding, while accomplishing a safe return is surely a consideration, the goal of the trip will be found in the manner of riding rather than the destination (Broughton 2006).

The idea of pursuing a goal for pleasure may be related to lack, or removal, of excitement in other areas of a person's life, and examples of this can be seen with the Inuit where:

> Young people who can no longer experience the excitement of hunting seal and trapping bear turn to the automobile as a tool for escaping boredom and focusing on a purposeful goal. (Csikszentmihalyi 2000: p. 71)

Many people may feel that in a sanitised, safe modern world, a similar experience may be sought in riding.

Routes into Riding

For someone to decide that they may want to ride firstly they must be aware of riding as an activity (McDonald-Walker 2000). From this they can progress into becoming a motorcycle rider, although there may be barriers that need to be negotiated first, such as hostility from family and friends who hold the opinion that biking is not appropriate because they view it as very unsafe or because of the way they view the biking community.

Family and Friends

The way most riders became aware of riding as a potential activity is through friends and family who already ride, and in this situation the hostility to riding may be lower, or even non-existent, and they may even have been encouraged to ride. Others may have become aware of riding via the media and, if the images projected are positive, this may encourage them to investigate taking up motorcycling. An example of this would be the effect that Ewan McGreggor and Charlie Boreman have had with their two televised epic trips, '*The Long Way Round*' and '*The Long Way Down*'. This has had the effect of attracting some into riding as well as increasing sales of the BMW type bikes they used on the trips (Telegraph Media Group 2007).

Image and Identity

Other motivating factors to attract potential riders may be the feeling that biking is fun or exciting, again positive images from the media and other riders may act as a catalyst for this. Even the outlaw image of 'bikers' may form an attractor for some potential riders. With modern society eroding our sense of belonging (Bradley 1996) people strive to find some identity and belonging (Hall 1989) and some may seek this through becoming part of the motorcycling community. Gutkind (2008) theorised that some may take up riding because of fear:

> … fear turns men to two wheels. Fear of getting too old. Fear of losing attractiveness. Fear of people higher up. Fear of a world too complicated. Fear of a world never seen. Fear of the ghosts that haunt at night. Riding a motorcycle eases the fear for some men. (Gutkind 2008: p. 204)

Convenience and Functionality

The practicality of biking is also an attraction with some who take it up for this reason finding that they enjoy riding and then moving on recreational biking. The congestion charge within London has made some drivers consider moving to a two-wheeled vehicle to avoid the charge (Smith 2007), with the added benefit that they are able to cut through the traffic and find it easier to park. Some companies, through their travel plans, encourage employees to ride to work and this in turn may encourage some into riding. The Institute of Highway Incorporated Engineers (2005) summarised some of the schemes including one at Vodafone whose:

> Employees are given an allowance of £85 per month to ride to work, and undercover parking spaces are provided for them. Changing facilities and lockers are also available to employees, along with the benefit of a motorcycle user group.

For a 16-year-old school leaver who needs to get to a place of work or education a moped may be the only viable option thus forcing them down this route, and from this acorn a rider may grow who continues to ride for many years. One rider who got into riding via this route commented:

> I needed to get to my place of work, only being sixteen at the time I had no option but to get a moped. I enjoyed riding so much that when I turned seventeen I bought a bigger bike; despite no one else in the family having ever rode before – I was hooked.

The part that motorcycles can play in social inclusion is demonstrated with the wheels to work scheme that provides small motorcycles to aid people getting to work. The scheme is described as:

> An incentive scheme designed to provide transport solutions to those who experience barriers to employment because of poor public or private transport. (The Countryside Agency 2002)

Whatever the reason for someone first riding a PTW, they become 'bikers' because they want to, with the majority remaining riders because they enjoy it. The riding of a motorcycle may not merely be an activity, but also a mode of being (McDonald-Walker 2000).

The discussion until now has mainly centred on trip purpose, reasons for ridings and the main things that are liked, or disliked, about riding. Although it is evident that enjoyment is a major factor in the use of PTWs, what kind of enjoyment is it and how is it derived?

Riding and Flow

Most bikes are ridden predominantly for pleasure rather than functionality and even when there is a functional reason for the ride, such as the convenience of commuting to work, there is still often an expressive or enjoyment element within the ride. As most riders seek enjoyment from their riding, with some trips being totally expressive, there must be elements within the riding task that generates this pleasure. Motorcycling is viewed by many as a risky activity (DfT 2006a; RoSPA 2001; Sexton et al. 2006) but because it is also an enjoyable activity some have drawn the conclusion that there must be some relationship between feeling at risk and the feeling of enjoyment within the riding experience. This raises the question 'do riders ride because of the risk, or despite of it?'

Flow

The aspects identified by the PTW users concerning their riding, all point to riding being a leisure activity for the majority of journeys undertaken. In most cases it is an activity undertaken through choice, in a person's free time and not undertaken as a commitment for work or for others, thus fulfilling the key elements to make it definable as a leisure activity (Torkildsen 2005). PTW use can seen as an 'active' leisure activity. Active leisure is more likely to result in longer-term enjoyment than passive activities such as watching TV. A sense of achievement and the psychological benefits associated with matching skills to challenge can produce life-enhancing experiences (Leitner and Leitner 2004). A situation that is likely to allow riders to find enjoyment is through experiencing 'flow'.

Csikszentmihalyi's Theory of Flow

Csikszentmihalyi's Theory of Flow suggests that when a person with a 'High Skill Level' is faced with a 'High Challenge' then this person can enter into a state he refers to as 'Flow'. Csikszentmihalyi describes this state as:

> The Holistic Sensation that people feel when they act with total involvement. (Csikszentmihalyi 2000: p. 36)

When someone is in the flow state they are so focused on the task that there is no attention left over to think about anything irrelevant or to worry about problems. Flow is an almost automatic, effortless, yet highly focused state of consciousness. People who have experienced flow often report nine aspects (Csikszentmihalyi 1990):

1. clear goals;
2. unambiguous and immediate feedback;
3. skills that just match challenges;
4. merging of action and awareness;
5. centring of attention on a limited stimulus field;
6. a sense of potential control;
7. a loss of self-consciousness;
8. an altered sense of time;
9. an autotelic (intrinsically rewarding) experience.

The Theory of Flow describes the four states of Apathy, Boredom, Anxiety and Flow. When the skill level and the challenge is low then an apathetic state is entered into, however, if the level of skill is higher than the level of challenge then boredom is the result; conversely when the skill level does not meet the challenge then anxiety exists. The flow state can be entered into when a person's skills, provided that they are not of a low level, are matched by the challenge that they are facing. Also for the flow state to be entered into not only must an individual's skills be matched to the challenges, but these challenges, and the skills needed to confront them, must exceed the normal levels of daily occurrence (Csikszentmihalyi and Csikszentmihalyi 1988). So the flow state can only be entered into when the challenges and skills are matched and are above the normal.

Just having a skill set of the correct level with respect to the challenge faced is not sufficient for a person to enter a flow state. A person must also have clear goals and instant feedback (Csikszentmihalyi 2000). Certainly within the riding environment instant feedback is available to the rider in the form of sensory input. This feedback will also provide information in a manner that gives enjoyment, which is the main aim of the ride. When a person is in a state of flow they are not happy, as to be happy a person would have to focus on the inner thoughts of being happy, and this would inhibit the flow state as it would divert some attention from the task. However, an enjoyable state of happiness may follow a flow experience. The flow state cannot be entered into when a person is passive, that is they have very little mental focus, so a rider must be completely focused on and engaged with the task.

Csikszentmihalyi (1997) also comments that flow activities are often bounded by strict rules, and these rules help to define what the activity is. Certainly this can be said to be true for driving and riding, with rules of the road and rules of physics needing to be observed. Driving is specifically mentioned by Csikszentmihalyi (1997) as an activity that can be 'flow related', which in turn may explain why it could be difficult to tempt some drivers out of their cars or away from the activity of riding.

An extension of the basic four channel model with an eight channel model is shown in Figure 6.1 (Csikszentmihalyi 1997; Massimini and Carli 1988)

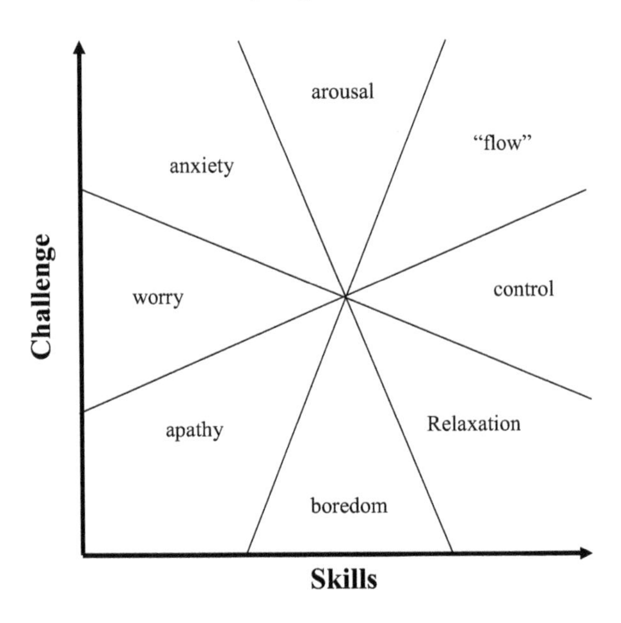

Figure 6.1 The eight channel model of flow

Within this model eight states are proposed, with a person's state depending on how matched their skill set is with the challenge faced. The activity of riding can be assessed in relation to this model to develop a clearer understanding of what happens to the rider when undertaking the experience. For example, if a rider is riding slightly outside their skill set limits then they may feel aroused, where they will be focused and involved but this state will not be an 'in control' or positive condition. However, if the challenge is slightly reduced then a rider could find himself in the area that can generate flow. Equally when a rider's skills slightly outstrip the riding challenge faced, then the rider is in control. As with the arousal condition the rider will be focused and involved, but this time they will be in control and the experience will probably be positive. When in this condition, if the challenge is increased by the rider riding a bit harder or faster then the flow state may be entered; however, if the rider pushes too hard then the arousal or anxious condition may well be entered.

While flow is basically matching a skill set to a challenge, there are traits and circumstance that can act as an inhibitor for achieving a flow experience. For example, a person who is excessively self-conscious would be unlikely to experience flow (Csikszentmihalyi 2000). This may be of particular importance for learners and inexperienced riders. It also may be an inhibiting factor for group riding where a rider may be concerned about being 'judged' by other riders. Conversely certain situations or activities can be an enhancer for flow achievement. For example, activities that are rhythmic, such as dancing, can help induce a state of flow (Csikszentmihalyi 1990). This may be of importance to riding as when negotiating a set of corners there is often a rhythmic flowing from leaning from one side to the other.

An understanding of what is happening at the neurocognitive level during the flow state also helps to comprehend some of the aspects of riding. On this subject Dietrich (2004) comments that:

> A necessary prerequisite to the experience of flow is a state of transient hypofrontality that enables the temporary suppression of the analytical and meta-conscious capacities of the explicit system. (Dietrich 2004: p. 746)

For the flow state to be entered the explicit cognitive systems must be temporarily suppressed. Therefore the brain must be running completely on the implicit system in a fully automatic mode where there is no mental processing power remaining to carry out other activities, such as daydreaming or analysing the task that is being undertaken. This is in agreement with the description of the flow state (Csikszentmihalyi and Csikszentmihalyi 1988) as being:

> An almost automatic, effortless, yet highly focused state of consciousness.

and that the:

> Task is performed, without strain or effort, to the best of the person's ability.

and that there is also:

> No sense of time or worry of failure.

Flow, therefore, is tied in with the automatic, implicit brain functions.

Implicit Memory

Implicit, or procedural, memory is not a memory area, rather a set of memory tasks (Graf and Schacter 1985) with these memories being skill- or experience-based; implicit skills have to be learnt via experience or training (Haberlandt 1999). These unconscious memory tasks contrast with the explicit memory tasks that a person is consciously aware of and can be expressed verbally. The two memory types are summarised in Table 6.4.

Implicit memory is related to 'rules of action', that is, how a person reacts given a set of circumstances with these actions becoming automatic with repetition. Frequently a person will have no awareness of how an implicit skill or action was learned (Allard 2001; Thorndike and Rock 1934). Broadbent (1958) carried out experiments that showed that a person could learn to do a task but could not verbalise how that task was carried out, rather all they could verbalise was what was verbalised to them in the way of instructions. For example, if a rider was

Table 6.4 Comparisons of memory types

Explicit Memory	Implicit Memory
Expressed by verbal communication	Not verbalisable
Conscious awareness	Inaccessible to conscious awareness
Flexible	Lacks flexibility
Slow	Fast

asked to describe how they changed gear they would most likely repeat how they were taught but not the actual actions they undertook.

As riders become more experienced many of the riding tasks will become proceduralised such as moving off and changing gears. This means that these actions will no longer require conscious thought of the slower explicit memory; these actions will become automatic and require little or no conscious thought. This can allow the rider to respond quickly to changing circumstances and enjoy the experience of the ride and the environment.

The mastery of skills can lead to enjoyment, but how does this link to risk for PTW users? In order to more fully appreciate the links between risk and enjoyment it would be desirable to have a controlled environment where all riders faced a similar group of circumstances where the actual riding experience can be accessed without disturbance from other road users and unusual hazards, for example, cars coming out of junctions. This would be difficult to achieve on a public road therefore alternative approaches to gathering this data is necessary. The opportunity of such near perfect conditions is possible by observing track days and gaining data from participants.

The Relationship between Risks and Enjoyment

'Track days' are when a track is turned over for use by the general public riding their own machines, however, this is not a 'free-for-all' use of the track. A track day is not a race situation, but an opportunity to test skills off the public highway. This type of event is the closest to a controlled experimental situation where ordinary bikers, often on their usual road bike, will all have a similar experience centred on their riding.

Broughton used the track in Edzell, Angus in Scotland as this track has clearly definable features that could be mapped and easily indicated to riders. Figure 6.2 shows the layout of the track. Broughton spoke to riders who were taking part in a motorcycle 'track day' (Broughton 2006; Broughton 2007; Broughton and Stradling 2005).

Figure 6.2 Edzell Track

With the cooperation of the track day organiser, riders were questioned on their feelings of risk, enjoyment, excitement and how much concentration they felt using each part of the 10 sections of the track. Riders were shown a map and asked to indicate where they felt most risk, experienced most enjoyment, used most concentration and felt most excitement. As task difficulty and task demand are potentially a key of riding in a similar way to driving (Fuller 2005), a selection of riders were also asked how difficult each area of the track was to ride and from this task difficulty ratings were created for each track section. Table 6.5 shows how many riders rated each section of track for maximum risk, enjoyment, concentration and excitement.

An analysis of the data showed that where riders stated that they had to concentrate hard, that they also felt at risk. One of the findings from the analysis of the dataset collected from the Edzell riders was that areas assessed as risky were also rated low for enjoyment, and conversely the sections that were rated as highly enjoyable were not rated as risky. Therefore, Broughton (2005) concluded that there was not an obvious linear link between risk and enjoyment. This conclusion was further demonstrated as only 13 per cent of the riders reported their highest levels of risk in the same section as their highest levels of enjoyment.

If risk does not lead to enjoyment, what are the factors that create enjoyment? In order to develop a deeper understanding of the processes that were taking place on the track, data on the way that each section of the track was perceived by riders was collected. Riders were asked to rate each section of track in terms of difficulty and give the reasoning for their assessment.

Table 6.5 Profile of track sections

Section	Risk	Enjoyment	Concentration	Excitement
1 – Straight	4%	22%	9%	13%
2 – Hairpin	65%	9%	57%	9%
3 – Chicane	4%	35%	13%	22%
4 – Curve	9%	39%	9%	35%
5 – Straight	0%	9%	9%	0%
6 – Hairpin	26%	9%	17%	4%
7 – Chicane	4%	43%	22%	26%
8 – Bend	9%	13%	9%	17%
9 – Straight	0%	13%	9%	0%
10 – Hairpin	30%	9%	39%	4%

Task Difficulty

Using this data, the 10 sections of the track were classified into five levels of task difficulty from having very low levels of difficulty to very high levels of difficulty (Table 6.6). The straights are rated as having low task difficulty because, despite being high-speed sectors, the level of skill needed to ride in a straight line is low and therefore little thought about riding line, or other features, is needed. The single corner is the next step up from a straight road, it is just a straight road with a single deviation in it, and therefore the task difficulty is slightly higher than for the straights. As chicanes comprise a series of corners that alternate between left and right their task difficulty is rated as medium; higher than for the corners. The hairpins on the track are rated as having the highest task difficulty as each hairpin has a fast section in its approach, therefore heavy braking is required before it is negotiated. Any braking on a PTW increases bike instability, heavy braking more so and during this manoeuvre the rider has to consider not only the riding line of the corner, but also where to brake, change gear and the position of other riders as hairpins are often a bottle neck for riders. The hairpin at the end of the main straight has a higher task difficulty rating compared to the other hairpins as the rider will be going very close to flat out when approaching it.

Broughton (2005, 2007) sought to establish the way that risk, enjoyment, concentration and excitement interact with task difficulty. He found that each had a distinctive relationship. Analysis of these relationships can help us to gain an insight into ways in which riders find enjoyment and how it links to risk.

Table 6.6 Track sections and task difficulty

Task Difficulty	Track Section
1 Very Low	1 – Straight 5 – Straight 9 – Straight
2 Low	4 – Corner 8 – Corner
3 Medium	3 – Chicane 7 – Double Chicane
4 High	6 – Hairpin 10 – Hairpin
5 Very High	2 – Hairpin

The earlier discussion on the theory of flow would suggest that for an activity to generate a flow experience, then it would have to present some challenge to the rider but not place the rider into a situation where there is a feeling of being out of control or their skills not being sufficiently honed to deal with circumstances. This would seem to be apparent when examining the data on enjoyment and the linked emotion of excitement against the task difficulty for each of the sections.

Task Difficulty, Enjoyment and Excitement

Figure 6.3 shows that as task difficulty increases so does enjoyment, however, when a threshold of task difficulty is reached enjoyment drops off. Excitement has a very similar relationship with task difficulty.

The peak enjoyment at medium levels of task difficulty demonstrates that high task difficulty situations for the average rider does not cause enjoyment, however, as enjoyment is also low in situations that the rider believes has low levels of task difficulty then some level of difficulty must be present for maximum enjoyment. Maximum enjoyment falls, to echo Csikszentmihalyi, between anxiety and boredom. Therefore some level of challenge is needed for the rider to find enjoyment in their riding. As there was no linear link between risk and enjoyment

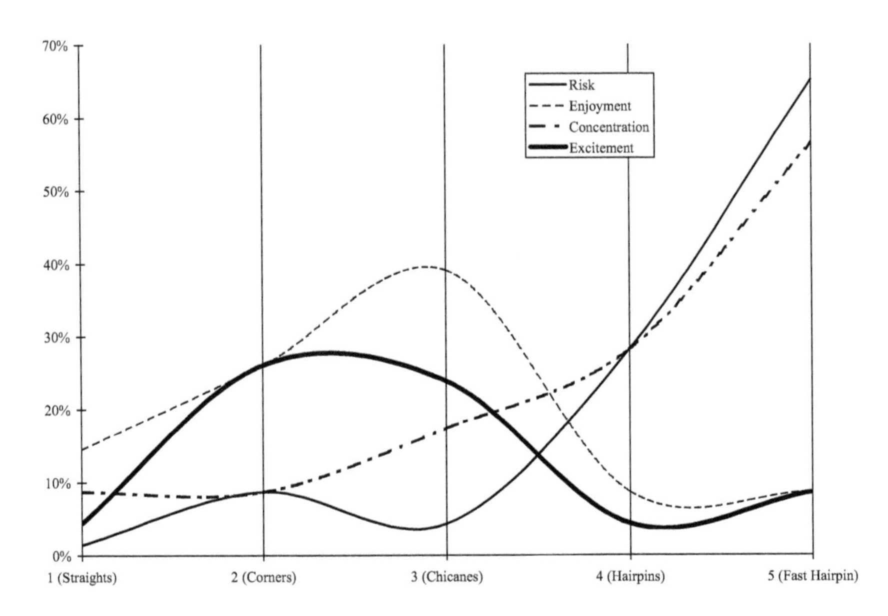

Figure 6.3 Task difficulty, risk, enjoyment, concentration and excitement

but a level of task difficulty is required for enjoyment to be found, how does risk link to levels of task difficulty?

Task Difficulty, Risk and Concentration

Risk is positively related to task difficulty as shown in Figure 6.4 (n = 28; r = 0.88); concentration is also closely related to task difficulty (n = 28, r = 0.91).

From these two correlations it would be predicable that there would be a strong relationship between felt risk and concentration, which is what Broughton (2005) found (n = 28; r = 0.98). Thus as a rider's perceived risk increases so also do the levels of concentration. As a situation becomes harder to ride in, demonstrated with an increase in task difficulty, both the level of felt risk along with concentration levels increases (Broughton and Stradling 2005).

These relationships between task difficulty, enjoyment, excitement, risk and concentration can be described in terms of Csikszentmihalyi's (1990) Theory of Flow.

Flow and the Edzell Riders

As discussed earlier, flow is a state that can generate enjoyment and happiness:

> The Holistic Sensation that people feel when they act with total involvement. (Csikszentmihalyi 2000: p. 36)

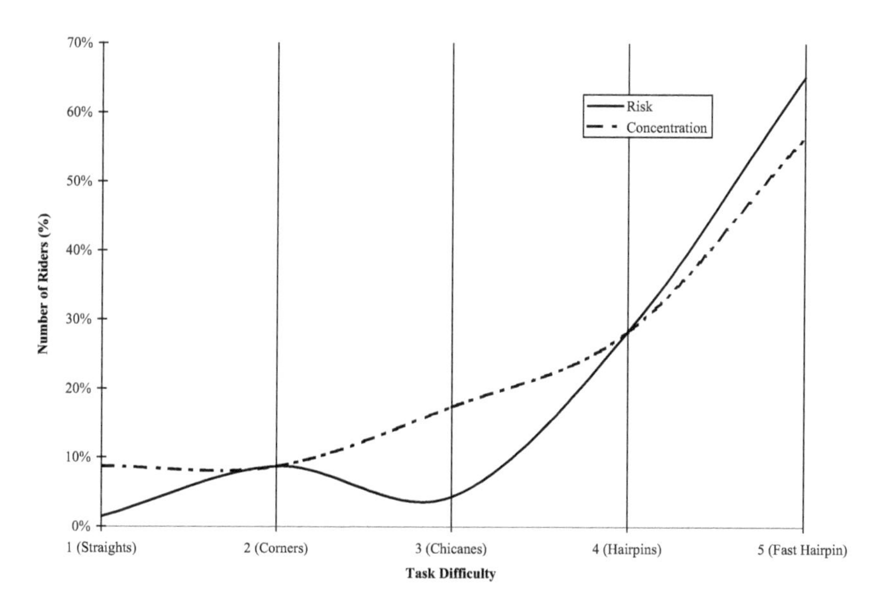

Figure 6.4 Task difficulty, risk and concentration

This total involvement is achieved by the matching of a person's skills at a task with the level of skill needed to carry out that task. When a person with a low skill set is faced with low challenge, then apathy is likely to be the resultant state. When the challenge outstrips the skill set, then anxiety is often the outcome. Conversely, with a high level skill set and a low level of challenge then a boredom state will probably be entered into. A modified model of flow has been developed that takes into account rider risk and enjoyment (Broughton 2008a). This model, shown in Figure 6.5, is a linear model that indicates the way that a rider progresses through the different states as task difficulty increases. Broughton (2007) describes this model:

> When task difficulty is low, boredom results, as task difficulty increases then the state of enjoyment is passed through and on to anxiety. The change of states is not instantaneous, that is one does not go directly from boredom to enjoyment, rather the boundaries are fuzzy. At the peak of enjoyment, just before the anxiety state begins, is the flow state.

This theory fits what was found in the Edzell data; at low levels of task difficulty there is a low level of risk and enjoyment, which is a state of boredom. However, as the task difficulty increases then the level of boredom also decreases. Enjoyment increases until a point is reached where the rider's skill level is matched to the challenge faced and the flow state is entered. However, if task difficulty continues to increase, then the rider's skill-set will not be up to the task and the flow state

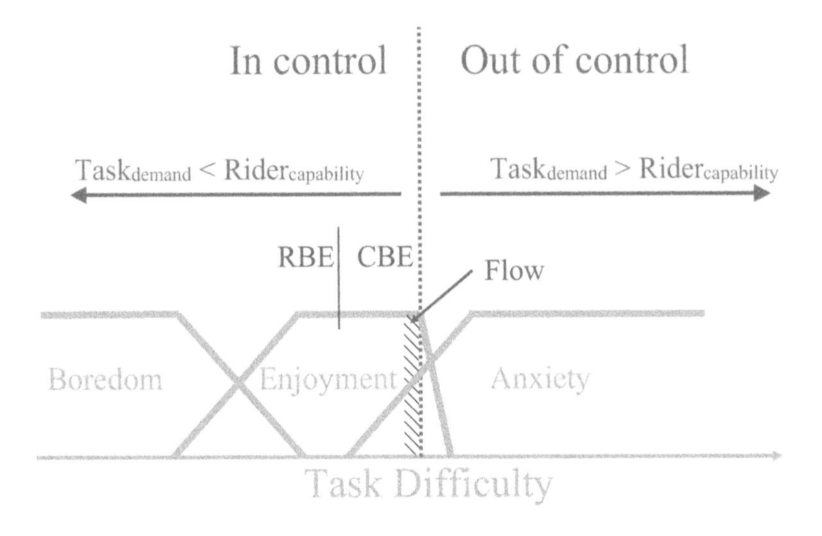

Figure 6.5 Linear model of task difficulty

will no longer exist. The rider's state will move from flow and into anxiety; this anxiety is felt as 'being at risk'.

Risk and Enjoyment

The results from data analysis of the track day experiment suggest that enjoyment is not linearly related to risk, but rather it comes from a moderate level of matching of skills with task difficulty. One rider stated:

> Satisfaction of control and use of skill. (Broughton 2007)

Therefore, riders seek challenge, but they do not want to put themselves in overly risky situations. However, in seeking challenge, they may find themselves in risky situations. When the task demand on riders approaches the limit of, or outstrips, their skill set then anxiety is felt, manifesting itself as feeling at risk. It could be summarised that PTW riders ride in spite of the risk, rather than because of it. This theory is based on data collected in the very specific and uncommon environment of a racetrack on a track day. The track day is dedicated to allowing riders to ride in virtually ideal conditions with few potential hazards such as poor road conditions or other road users with poor understanding of bikers, but how does enjoyment and risk relate in the less idealistic settings mainly experienced by PTW users? The following chapter develops this theme and examines this relationship between risk and enjoyment in more detail.

Chapter 7
Enjoyment and Risk

Introduction

The relationship between risk and enjoyment was explored in the previous chapter where is was suggested that while there is a relationship between risk and enjoyment, enjoyment is not a result of risk and risk may even have a suppressive effect on enjoyment. As found in previous studies relating to risky activities in the realm of sport (O'Sullivan, Zuckerman and Kraft 1998; Scanlan and Simons 1992; Slanger and Rudestam 1997), it would seem that the majority of riders ride despite the risk, not because of it. Rather it is the challenge of matching skills to the situation that is the appeal. However, the dataset that Broughton used to draw these conclusions was gathered in the controlled, but artificial, situation of a racetrack. It is also possible that only a particular type of rider would participate in events such as track days. This chapter seeks to develop a deeper understanding of risk and enjoyment in the more complex riding situations faced on the public roads.

Testing on the Real Road

Testing risk and enjoyment interaction in a real road situation would prove difficult due to ethical and safety issues, which must be considered with any 'on road' experiment (Hutchingson 2008). Therefore 'on road' data was collected by asking riders to consider photographic scenarios. The use of pictorial real-life riding situations allowed data to be collected in a safe and ethical manner but with the additional benefit of reliability of scenario data as they are presented in a systematic manner so that all riders are faced with exactly the same situation to assess. This would not be possible in a true 'on road' experiment as environmental factors would influence the results, for example, the weather or a dog running out into the road.

The Six Scenarios

Broughton used six photographic scenarios that were shown to Powered Two-Wheeler (PTW) riders (Broughton 2007); these are shown in Figure 7.1. Each of the photographs was assessed by the motorcyclist for riding enjoyment and perceived risk, riders were also asked for the reasons why they assessed each road scenario as they did.

Figure 7.1 Scenario pictures (courtesy of Owl Research Ltd)

Scenario one is a long straight road that mimics the straights at the Edzell racetrack; however, there is also a junction on the right, and access from a field to the left. There are three drain covers on the right hand side of the road, surrounded by repair work with a tar seam (over-banding). Such features are common elements of public roads.

Scenario two was selected because of the sweeping right hand bend, with the bend going under a bridge followed by left-hand corner in the distance. The road under the bridge is in shadow. There are national speed limit signs visible so there is no ambiguity regarding the maximum legal speed. In the distance, past the bridge, there is a triangular warning sign.

Scenario three is an urban scene showing the approach to a busy roundabout. There is a green coloured bus lane on the left hand side. In the foreground there is a drain on the boundary of the road and the pavement, and also grooves in the road towards the outside of the lane. Oil may be deposited within the bus lane. There are four other roads converging on the roundabout, with vision being partly obscured by shrub and tree growth within its centre. A lorry carrying a skip is waiting to enter the roundabout, while on the roundabout there is a red car, followed by a black one, with a vehicle joining the roundabout from the right. On the approach the car within the bus lane is braking as it gets closer to the skip lorry, there is also a vehicle in the outside lane. A pedestrian, who is crossing the road, is waiting on an island in the middle of the carriageway.

Scenario four was selected because it is an open road with a significant amount of traffic using it. In the mid-distance there is signage indicating a garage, beyond the garage there is a shaded out section in the centre of the road. A car is directly in front, with a line of three more vehicles some distance ahead of that. Four cars are approaching in the immediate vicinity, and at least one of these has its front lights on.

Scenario five is another urban setting and was selected because it is obviously urban and has shops lining the road. There is no moving traffic present, although there are some parked cars and four pedestrians, two of whom are walking on the road carrying shopping. In the mid-distance a junction emerges from the left and in the foreground the road shows signs of repair.

The final scenario, six, was selected because it is a rural road with a sweeping right hand corner. The centre of the road is marked with double white lines. The road is lined with trees that are losing their leaves and these have been deposited on the roadside. On the inside of the bend there is a pavement bounded by a wall and there is a grass verge on the outside of the corner. A carrier bag is lying on the pavement.

Broughton chose these six selected scenarios as they give a reasonably representative cross-section of different road conditions, layouts and potential hazards. Rider assessment of these scenarios was then used as a basis for analysing attitudes and perceptions in a diversity of situations.

The Rating of Factors

Prior to the main research, a group of experienced riders were asked to assess each scenario in terms of risk and enjoyment using a five-point scale, and to give reasons for their assessment. Broughton used these reasons to identify six risk factors and five enjoyment factors (Broughton 2007; Miller and Crabtree 1992); these factors are illustrated with example quotes below.

Risk Factors

- Road surface quality

  Surface looks uneven (bumpy), smooth patches where tar has worked up indicates heavy use, manhole covers staggered and is potentially dangerous to bikers in a emergency situation such as heavy braking. (Scenario one)

- Risk caused by road features, such as road size, roadside objects, junctions, and so on.

  Possibly traffic emerging from side roads and farm tracks. (Scenario one)

- Level of visibility

  Quiet road, but with a bend that prevents a view into the distance. (Scenario two)

- Likelihood of a rider distraction

  Slow, busy, stop and go, with lots of distractions. (Scenario three)

- Risk presented by other road users (including pedestrians)

  Other road users not signalling, cars taking up other lane (sneaking in) … and cars pulling out on me. (Scenario three)

- Riding in an enthusiastic manner (temptation)

  Very straight therefore temptation to go too fast. (Scenario one)

Enjoyment Factors

- Surroundings, scenery, etc.

  I would enjoy this road because I like to travel in the country and look at the crops and animals. (Scenario one)

- Challenge

  Challenging curve but limited line of sight. (Scenario six)

- Bends

Negotiating the bends and it being a country road. (Scenario two)

- Speed of riding

Chance to open the throttle. (Scenario one)

- Overtaking opportunities

Yes there's traffic but we can get some good overtakes in. (Scenario four)

Five of the six risk factors, such as road quality and other traffic, are third party with only one, 'Riding in an enthusiastic manner' that can be directly attributable to the rider's internal state. Four of the five enjoyment factors are directly related to the riding activity with one factor, the surroundings, being external to riding. An obvious difference between bikes and cars is that in most cars the driver is surrounded by a metal and glass cage while on a bike the rider is more exposed and therefore in a better position to experience the 'great outdoors'.

Profiling Risk and Enjoyment

One purpose of using the scenario picture-based research method is to evaluate how risk and enjoyment relate to each other when riders are riding in a 'real' situation on the public road as opposed to in the racetrack conditions, as reported in Chapter 6.

Riders were asked to evaluate each of the photographic stimuli for riding risk and riding enjoyment on a five-point Likert scale with one for very low and five for very high. The dataset from the responses allowed for the mean risk and enjoyment to be calculated for each scenario, and these are shown in Table 7.1. For scenarios

Table 7.1 Means of risk and enjoyment rating by scenario

	Risk	**Enjoyment**
Scenario one	2.66	3.09
Scenario two	3.22	3.18
Scenario three	3.70	2.31
Scenario four	2.56	3.33
Scenario five	3.70	2.05
Scenario six	2.89	4.17
Overall	3.75	3.61

of high risk, such as scenario three, the level of enjoyment is low, while for a high enjoyment scenario, such as number six, the risk is assessed as medium.

There is very little difference in the distribution profiles of risk and enjoyment ratings over all scenarios (Figure 7.2) and it therefore might be concluded that risk positively correlates with enjoyment with the middle risk and enjoyment ratings being the peak value of risk and enjoyment. However, consideration of risk and enjoyment for each scenario shows this not to be the case, each scenario is now discussed in reverse order of risk, from highest risk to lowest.

The urban scenario five has a very high-risk rating (mean of 3.70), coupled with a low enjoyment rating (mean of 2.05). Within this urban area the main reasons given for feeling at risk were other traffic, including pedestrians and the quality of the road surface, these aspects are well captured by the comments of one rider who said:

> Pedestrians and shoppers spell risk of people thinking of things other than traffic around them. Cars are less likely to see a bike coming down the road. People may have kids that run out without warning. Road surface liable to be dodgy at best, often with potholes etc. (Broughton 2007)

> AAAAhhhhhhhhhhhh PEDESTRIANS shopping ... No brains ... the buggers will be all over the place like zombies !!!!!!!! (Broughton 2007)

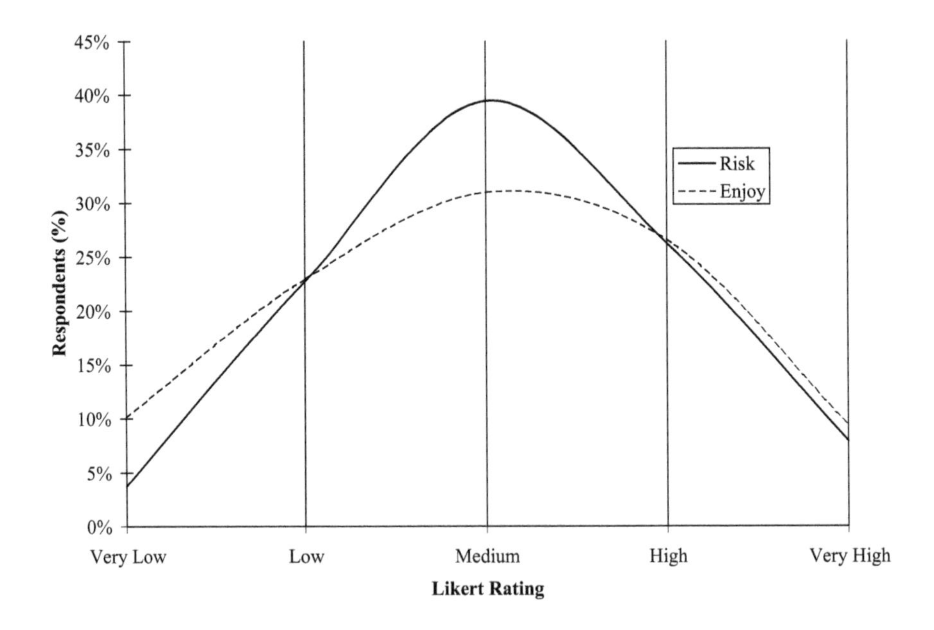

Figure 7.2 Risk and enjoyment (all scenarios)

There were very few comments that reflected a positive view on enjoying this scenario, rather most expressed that the level of felt risk prevented an enjoyable situation:

> I don't like driving in town because it requires extraordinary amounts of attention. There are too many circumstances that constantly change, which requires lots of mental processing. (Broughton 2007: p. 457)

> Low rewards for all the risk –in town riding can be very dull. (Broughton 2007)

There is a large difference between risk (mean of 3.70) and enjoyment (mean of 2.31) for scenario three, the busy urban roundabout. The main reason given for the high level of risk was other traffic (54 per cent), followed by road features (21 per cent):

> Primary danger is all the other traffic drivers about. Are they as alert or skilled as I am? Hard to predict what they might do next. All skill here is placed on surviving and negotiating the roundabout, not enjoying the fact I am on two wheels. The risk of diesel spill somewhere round here is massively increased. (Broughton 2007)

Only a few riders commented about the possibility of this scenario being enjoyable to ride and these comments were focused on the bends that roundabouts naturally generate, but this enjoyment would mainly be confined to when the roundabout was not busy:

> Dry, grippy, quiet roundabouts can be fun. (Broughton 2007)

Risk and enjoyment for scenario two, the right hand sweeping corner going under the bridge, are closely matched with a mean risk of 3.22 and mean enjoyment equating to 3.18. Over 60 per cent of the respondents' comments concerning risk were coded as poor or lack of visibility with bends (34 per cent), for example:

> Bridge is blocking visibility. (Broughton 2007)

> Cannot see what is coming towards me. Also, I may be less visible when in the shadow under the bridge. (Broughton 2007)

> Unsure whether road continues to bend right after bridge, or cuts left. Either direction, visual distance beyond the bridge is EXTREMELY limited, for me and for drivers approaching me. (Broughton 2007)

Pleasant scenery (10 per cent) was the most frequent responses for enjoyment:

> It would be a good road to take in the scenery. (Broughton 2007)

> If the other traffic behaves, the circle would be fun. (Broughton 2007)

Scenario six is the right hand sweeping curve in a rural setting; this has a very high enjoyment rating (mean of 4.17) with a medium risk rating (mean of 2.89). This scenario was rated as the most enjoyable with bends being most often given as the reason for enjoyment (42 per cent):

> Everyone likes a good sweeper. (Broughton 2007)

> Twisty open rural road. Looks like fun to play on. (Broughton 2007)

> Empty bend fun, what bikes were made for. (Broughton 2007)

The main reason given why this scenario might generate a feeling of risk is the lack of visibility (33 per cent):

> This is a high risk piece of road simply because it could be taken at high speed and there is a blind corner. (Broughton 2007)

> I can't see what's round the corner, whatever's round the corner can't see me either. (Broughton 2007)

The mean risk and enjoyment for scenario one, the long straight road in a non-urban setting, are both below the mean of all the scenarios (risk = 2.66, enjoyment = 3.09). An examination of the reasons for risk and enjoyment for this scenario shows that nearly 50 per cent identified the road surface as a major cause of risk, followed by road features (28 per cent). These two elements were described by one rider:

> Surface appears poor. Possibly traffic emerging from side roads and farm tracks. (Broughton 2007)

When discussing enjoyment 26 per cent gave pleasant surroundings and 23 per cent speed as reasons. Despite the road being a long straight road that would be suited to very high speeds, only a quarter said that speed was an enjoyment factor with more commenting on the pleasant scenery, for example:

> Straight, ok for higher speeds. (Broughton 2007)

> Like riding in the country. Scenic, reduced traffic volume. (Broughton 2007)

In scenario four, the straight main road with some traffic on it, enjoyment (mean of 3.33) outstrips risk (mean of 2.56). Around 25 per cent said enjoyment could be found in speed with 19 per cent stating that overtaking could provide enjoyment:

> Chance to open the throttle. (Broughton 2007)

> Looks a bit boring but safe and progressive overtaking is a skill and can be fun. (Broughton 2007)

When commenting on risk with 35 per cent saying that other traffic was a cause of risk:

> Good clear view ahead lots of space for getting out of trouble if a car decides to do something silly. Biggest risk here is an oncoming car overtaking in face of traffic. (Broughton 2007)

Analysis of the data relating to risk and enjoyment for the six scenarios suggests there is something more complex than a linear correlation between risk and enjoyment. The interaction between risk and enjoyment varied with each scenario and the reasons given for risk and enjoyment were also scenario specific. The comments show that riders tend to know what causes them to feel at risk and also the situations that can give rise to enjoyment. This suggests that at some psychological level riders know for themselves how risk and enjoyment interact.

The comments made by riders show that what gives rise to enjoyment for one rider may be seen as a risk generator for another rider, for example, consider the two contrasting rider quotes concerning scenario six:

> Blind curves, low visibility of on coming traffic or road hazards. (Broughton 2007)

> Gentle curve, decent visibility. (Broughton 2007)

As the perception of risk and enjoyment is not a constant across all riders, is there a pattern that can be used to classify how riders view risk and enjoyment?

Interaction between Risk and Enjoyment

Figure 7.3 shows the interaction between risk and enjoyment using mean value for each scenario and plotted in risk order. As risk increases so does enjoyment, until a peak of enjoyment is reached; as risk further increases enjoyment drops off rapidly.

Broughton and Stradling used specialist pattern recognition software (Broughton and Stradling 2005; Pao 1989) to categorise riders into risk/enjoyment groupings;

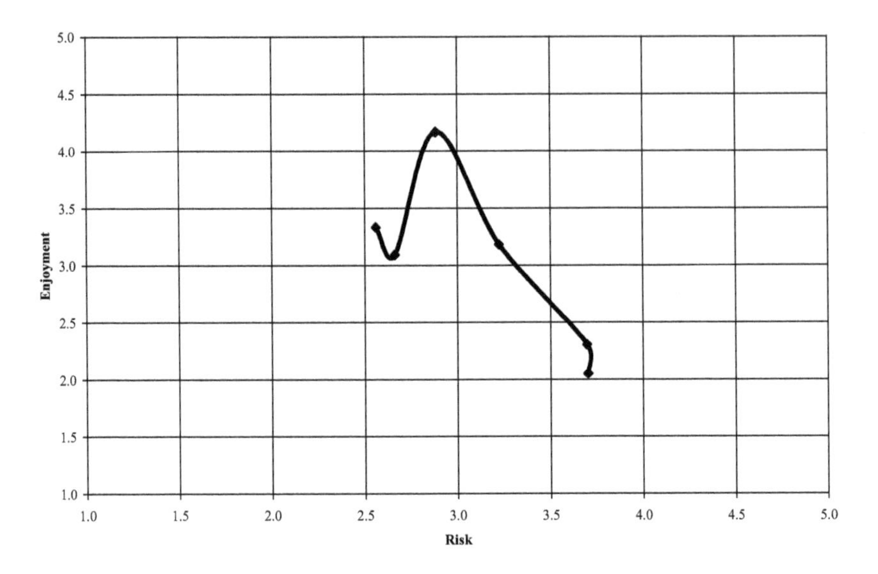

Figure 7.3 Risk against enjoyment

for more details on the method used see Broughton (2007). The pattern recognition software found three risk/enjoyment profiles. These were entitled 'risk acceptors' for those whom, as risk increases, so does enjoyment until a threshold point is reached, then enjoyment decreases as risk continues to increase; 'risk averse' for those for whom enjoyment decreases as risk increases; and 'risk seekers' for whom enjoyment increases as risk increases. The number of riders categorised into each group is shown in Table 7.2 (note that the software was unable to classify 2 per cent of the sample).

Risk Acceptors

The risk acceptors profile is very similar to the overall profile (Figure 7.4), which should not be surprising as this group makes up nearly half of the sample. Enjoyment increases with risk, until an enjoyment maximum is reached before rapidly declining as risk continues to increase.

The mean value of enjoyment and risk for risk acceptors was calculated, and then sorted into ascending risk order (Table 7.3).

Risk acceptors gain most enjoyment at mid risk (Figure 7.4). These riders are happy to accept a level of risk that enables them to enjoy their riding, but once this level has been exceeded then the activity becomes less enjoyable. Examining these data in terms of flow, it would seem that risk acceptors seek a middle way between boredom and anxiety.

Table 7.2 Risk type groupings

Risk Type	%
Risk acceptor	48%
Risk aversive	42%
Risk seeker	8%
Undetermined	2%

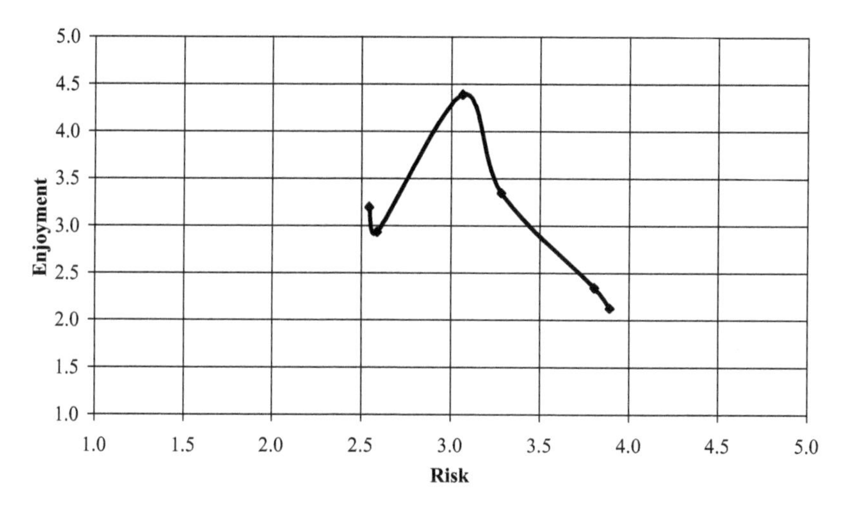

Figure 7.4 Enjoyment profile of risk acceptors

Table 7.3 Mean risk and enjoyment values for risk acceptors

Scenario	Risk	Enjoyment
4	2.54	3.20
1	2.59	2.93
6	3.07	4.39
2	3.28	3.35
3	3.80	2.35
5	3.89	2.13

Risk Averse

The risk averse profile is shown in Figure 7.5.

The risk averse profile tends towards a straight line with a negative slope, demonstrating that, for this group, enjoyment reduces as risk increases. The risk averse rider does not equate risk with enjoyment; as risk increases enjoyment decreases and their enjoyment is low whenever there is perceived risk.

The mean risk and enjoyment values were calculated and ordered by risk in the same way as for the risk acceptors (Table 7.4).

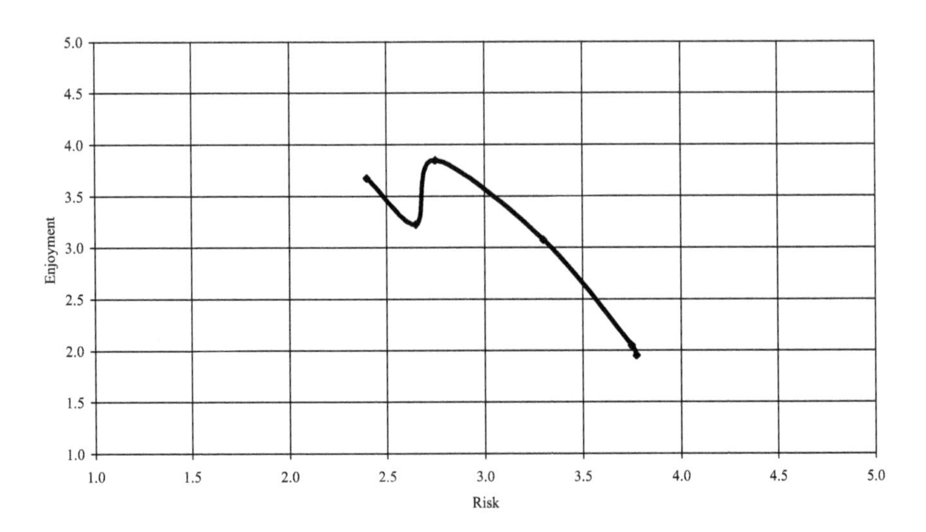

Figure 7.5 Enjoyment profile of risk averse

Table 7.4 Mean risk and enjoyment values for risk averse

Scenario	Risk	Enjoyment
4	2.40	3.68
1	2.65	3.23
6	2.75	3.85
2	3.30	3.08
3	3.75	2.05
5	3.78	1.95

Another characteristic of the risk averse group that becomes apparent from data collected by Broughton is that they less likely to brake hard (15 per cent said they would at least sometimes) compared to 75 per cent of risk acceptors and 95 per cent of risk seekers ($\chi^2(2\ df,\ n = 143) = 18.113,\ p = 0.020$). Despite the risk averse riders being less willing to brake hard, they feel that they are riding closer to their limits, riding at an average of 86.9 per cent of what they consider to be their maximum speed, with the other two groups riding at a level of 76.3 per cent ($t(130) = 2.218,\ p = 0.005$). This group is also more likely to accept extra training and they claim to enjoy riding more than the other groups.

Risk Seekers

Risk seekers are the opposite of risk averse; as risk increases, enjoyment increases therefore a positive correlation would be expected. The mean risk and enjoyment for risk seekers is shown in Table 7.5, ordered by risk as with the other groups.

For risk seekers, enjoyment dramatically increases with perceived risk (Figure 7.6). It may be that for this class of rider there would be a threshold where the risk becomes too high to give them enjoyment, but this threshold point is considerably higher than for the other rider types and was not captured in the set of six stimuli used here. This was by far the smallest group of riders (8 per cent of the sample). So while they may show tendencies toward behaving in the way one would expect from media portrayal bikers, they are very much in the minority and cannot be seen as 'typical'.

Table 7.5 Mean risk and enjoyment values for risk seekers

Scenario	Risk	Enjoy
4	2.63	1.90
1	3.00	2.40
6	3.13	3.38
2	3.25	3.60
3	3.38	3.88
5	3.50	4.05

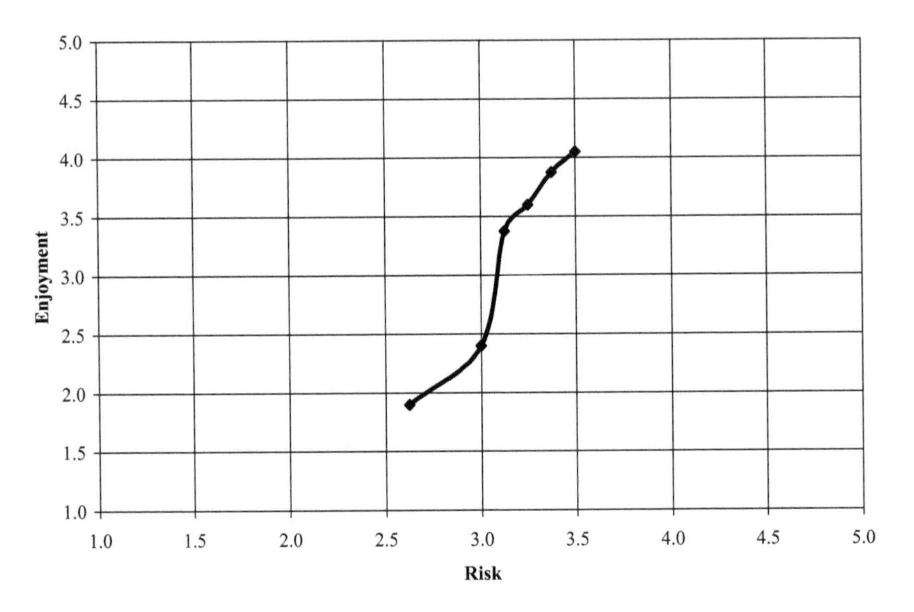

Figure 7.6 Enjoyment profile of risk seekers

The Relationship between Risk and Enjoyment

The relationship between risk and enjoyment in scenarios that riders face regularly on the public highways suggests that outwith the constrained and controlled environment presented in the track day situation, the relationship becomes even more complex. In the track day the riders had a fairly stable riding situation that allowed them to practise skills in a relatively unfettered situation that is only possible in such a controlled arena. In real road situations, such as presented in the picture scenarios, the environment is far more volatile. Those scenarios rated more highly for enjoyment tended to be those with fewer interactions with other road users, including pedestrians; such interactions would seem to increase risk felt without adding enjoyment. Given that other road users are often the cause of crashes for PTW users (ACEM 2004), the belief that such interactions increase risk is justified by the statistics.

Enjoyment is derived from the actual riding process: riding around bends and overtaking with the only external factor evident for enjoyment being the surroundings. This suggests that being in control is important to riders. This idea of being in control relates to task difficulty and ideas of flow discussed in Chapter 6. Riding enjoyment is greatly amplified when a rider feels able to ride expressively, matching the challenge of the situation with his skills and ability.

Although risk was a key element in the scenarios only one of the three rider groups were identified as actively seeking risk and they comprised only 8 per cent of the sample; this is in line with the 13 per cent of track riders who reported high enjoyment in areas that they felt that they were at high risk. The two main groups were quite evenly split between risk acceptors (48 per cent) and the risk averse (42 per cent). For risk acceptors, enjoyment increased as risk increased but then rapidly decreased after a threshold of acceptable risk had been reached. For the risk averse group, risk has an inverse relationship with enjoyment such that as risk increases, enjoyment decreases more rapidly. This is further evidence to suggest that while risk is an inherent part of riding, it is rarely sought as a means of increasing enjoyment.

Enjoyment and Risk Factors

The ways in which riders experience and perceive risk influences enjoyment but for most riders high risk is not a factor that leads to enjoyment. For the majority of riders risk was accepted as a factor in riding but it is not actively sought. Risk is perceived when levels of task difficulty exceeded the upper limit of the skill level of the rider, but enjoyment is maximised where task difficulty matches the rider's perceived skill; the boundary between these two states is small. In order to enable a full understanding of riders and their motivations a deeper understanding of the factors that influence risk and enjoyment are necessary. Previously in this chapter 11 elements were identified that either affected perceived risk or felt enjoyment:

1. road surface quality
2. risk caused by road features, such as road size, roadside objects, junctions
3. level of visibility
4. likelihood of the rider/driver being distracted
5. traffic (risk presented by other road users, including pedestrians)
6. temptation to ride in an enthusiastic manner
7. surroundings (scenery, etc.)
8. challenge
9. bends
10. speed of riding
11. overtaking opportunities.

In order to gain a clearer understanding of the relationship between these elements, perceived risk and felt enjoyment, the data reduction technique of factor analysis was used. This technique enables patterns in the dataset to be more easily identified.

Enjoyment and Risk Types

Factor analysis on these elements along with perceived risk and felt enjoyment produced three factors (for more information see Broughton 2007); two of these factors related to enjoyment and one to risk. By grouping the data in this way it makes it easier to see patterns and identify trends. These factors can then be examined in relation to individual elements to assess how they interact and thus allow more depth to the interpretation of the data.

Features of Enjoyment

The factor analysis showed that enjoyment can be found in two statistically separable, and thus plausibly psychologically separate, ways, with eight road and riding features relating to these two types of enjoyment (Broughton 2008a):

1. road surface quality
2. visibility
3. temptation to ride enthusiastically
4. the surroundings
5. challenge
6. bends
7. speed
8. opportunity for overtaking.

Each of these factors is examined with respect to enjoyment to establish what kind of relationship each has.

Enjoyment and Road Surface Quality

Enjoyment rises steadily with road surface quality; however, road surface quality does not feature as an element of risk within the factor analysis, despite the need of good traction for motorcycles. This lack of correlation with risk, coupled with its relationship with enjoyment, suggests that poor road surface quality is an enjoyment inhibitor rather than a risk enhancer.

Enjoyment and Visibility

When visibility is below a certain threshold then enjoyment is curtailed, however, once the visibility exceeds this threshold then enjoyment is relatively constant. This suggests that as visibility increases so does enjoyment, but once a certain level of visibility is reached then any further increase does not generate any further increase in enjoyment.

Enjoyment and Temptation

There is a slow rise in enjoyment as temptation increases. Therefore in areas where a rider might be tempted to ride in an enthusiastic manner enjoyment is found: enjoyment is found where there are opportunities to ride enthusiastically.

Enjoyment and Surroundings

Areas that have good surroundings are generally areas of high enjoyment; this may be because the surroundings give enjoyment as indicated by some of the quotes earlier, or it could be that other features that exist in areas of good surroundings give rise to enjoyment, or a combination of these factors.

Enjoyment and Challenge

Challenge is related to enjoyment but as with visibility, lack of challenge may be an inhibitor of enjoyment rather than a high level of challenge giving rise to enjoyment. However, for some riders, challenge may not be needed for enjoyment (Broughton 2007). The profile seen in Figure 7.7 shows how enjoyment changes with challenge. For a challenge rating of up to seven, enjoyment is relatively constant, however, once this challenge threshold is exceeded then enjoyment increases rapidly. Therefore to enhance their riding enjoyment riders may need to push themselves until they feel challenged. This challenge can be sought through the nature of the roads ridden or the speeds at which these roads are ridden.

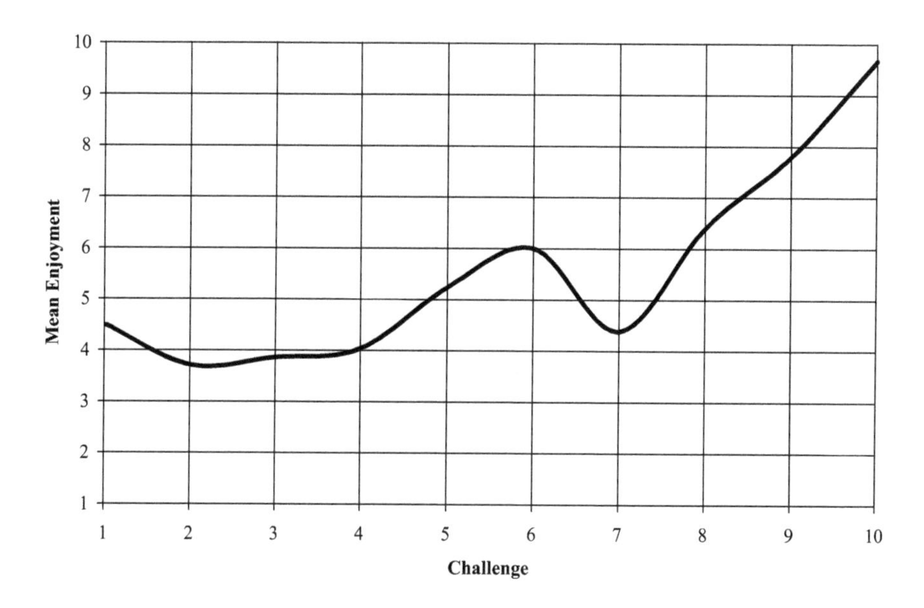

Figure 7.7 Challenge and mean enjoyment

Enjoyment and Bends

In scenarios where the road is straight then mean enjoyment tends towards being low, although around a third of riders claim to find high enjoyment on straight roads (Broughton 2006; Broughton 2008a). The percentage of riders who find enjoyment from straight roads correlates with those who say they find enjoyment in low challenging riding situations.

By plotting the mean levels of enjoyment against bends the data suggests that there is a positive correlation between bends and enjoyment levels; as the road becomes bendier, enjoyment increases, as shown in Figure 7.8. However, there are signs of the relationship being exponential, that is increasing at a quicker rate as the bendiness increases.

Enjoyment and Speed

The relationship with speed and enjoyment is an obvious and dramatic step function at a speed rating of five to six (Figure 7.9). Enjoyment is at a near constant level of around four until a threshold is reached, at a speed rating of six. At this threshold point, enjoyment rises and settles at a value of just over six. Therefore a level of speed is needed for enjoyment with riding at very low speeds not being conducive to enjoyment. It would seem then that speed is a modifier of challenge level and a degree of speed is needed for enjoyment. It can thus be deduced that a minimal level of challenge is needed for enjoyment.

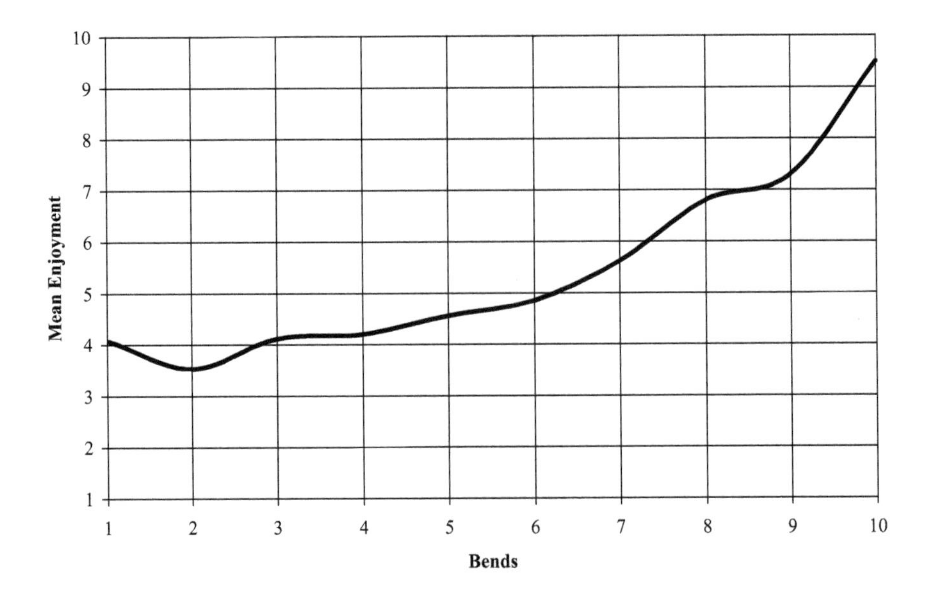

Figure 7.8 Challenge and mean enjoyment

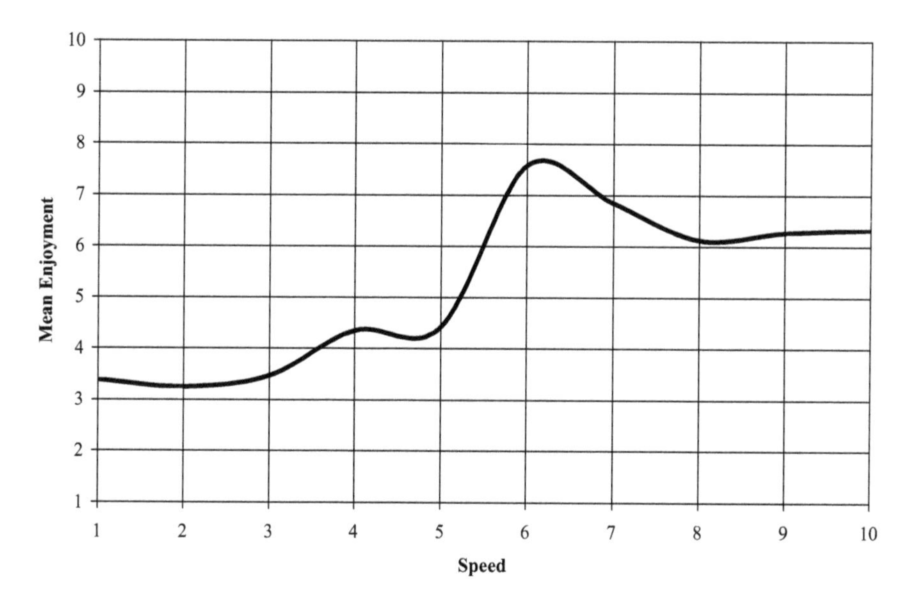

Figure 7.9 Speed and mean enjoyment

Enjoyment and Overtaking

There is also a step function in enjoyment when compared to the ratings of overtaking with this being very similar to enjoyment from speed, but the overall enjoyment is around 15 per cent lower for overtaking that for speed. This shows that, for some riders, speed enjoyment can be found where other traffic has to be negotiated; however, for other riders this traffic and the act of overtaking reduces enjoyment.

Enjoyment Factors

Two clear factors become evident that relate to enjoyment. Factor 1, with an enjoyment weighting of 0.48, is related to road surface quality, visibility, temptation, surroundings, speed and overtaking. This factor is concerned with the rider gaining enjoyment from riding fast on straight roads, hence the weighting of speed at 0.85. Bends are not related to this factor (-0.08). The inclusion of visibility and overtaking is because good visibility is a factor of most straight roads where it is easy to ride fast and overtake. This factor is linked to pure speed; therefore it is about getting an 'Adrenalin Rush' without having to push oneself skill-wise. This factor is designated by Broughton (2008a) as 'Rush Based Enjoyment' (RBE).

Factor 2 with an enjoyment weighting of 0.52 is related to road surface quality, temptation, surroundings, challenge and bends. This factor is about gaining enjoyment from the challenge of riding with bends providing that challenge. Both challenge and bends are heavily loaded on this factor; 0.88 and 0.90 respectively. As this factor is challenge-related this type of enjoyment can be seen as relating to the flow state (Csikszentmihalyi 1990) and was designated by Broughton (2008) as 'Challenge Based Enjoyment' (CBE).

Three common components are found in both factors: road surface, temptation and surroundings. Two of these elements, temptation and surroundings, are enjoyment enablers that are required for both types of enjoyment. Road surface quality is an enjoyment inhibitor as the lack of road quality prevents a ride being enjoyable.

The two enjoyment types can be compared to the difference between bungee jumping and rock climbing. Both of these activities can be enjoyable, yet enjoyment is found in completely different ways. Bungee jumping does not require much skill in throwing oneself off a high place with a piece of elastic saving one from death, yet this is very enjoyable for those who are seeking an adrenaline buzz. Rock climbing on the other hand is a sport where a climber pits their skills against the challenge presented by the rock face with enjoyment found in the skill/challenge match. Enjoyment can be found in either, or both, types of activity (Broughton 2006). Some riders may seek RBE, some may be more attracted by CBE. This may vary over time and according to the characteristics of the rider. The profiling information on the riders discussed earlier was used to assess how this impacted on the type of enjoyment sought.

Enjoyment Demographics

Those who ride the lower performance bikes were more likely to have 'Rush Based Enjoyment' experiences than those riding machines towards the upper end of the performance spectrum (Figure 7.10). For those experiencing high RBE, around 60 per cent ride either low, or very low, performance motorcycles.

As bike performance increases then mean enjoyment from rush decreases. This dichotomy may be due to rider experience as experienced riders are more likely to ride the higher performance machine. As these riders generally have a higher skill level this may mean that they are less likely to gain enjoyment simply from 'rush' and more likely to gain enjoyment from challenge.

This is further emphasised when rider age is considered; only 2 per cent of riders over 50 experience high levels of 'Rush Based Enjoyment', compared to 51 per cent of under 35s. Younger riders tend more towards 'Rush Based Enjoyment', with older riders more likely to be 'Challenge Based' (Broughton 2007), and as age restricts bike performance, this helps to explain the dichotomy of performance against speed.

The gender split shows that females are twice as likely to be in the upper ends of the 'Rush Based Enjoyment' category compared to males (14 per cent of males compared to 29 per cent of females). However, when gender is compared

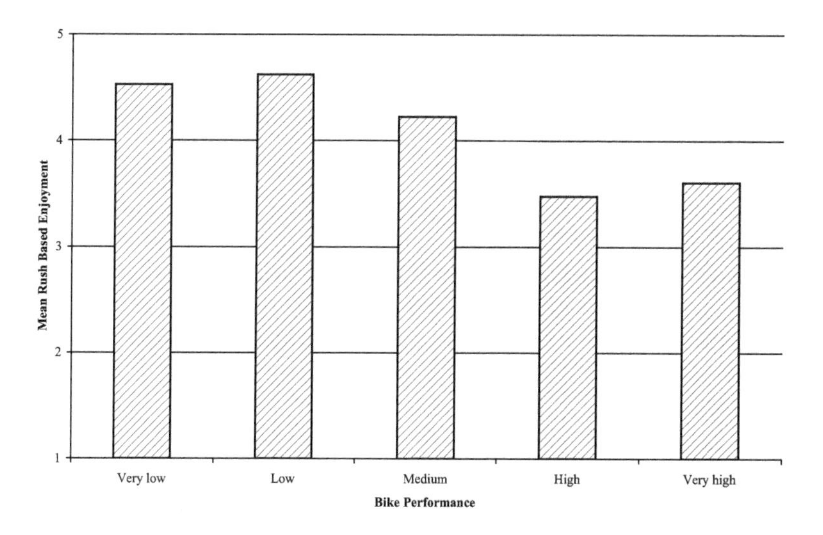

Figure 7.10 Mean rush based enjoyment and bike performance

to 'Enjoyment Types', it is noticeable that females are over represented at the extreme ends being more likely to be either high 'Challenge Based' or high 'Rush Based' while males are more likely to be within the middle groupings ('Slight Challenge' to 'Slight Rush'). Females make up the minority proportion of riders, and it is still less socially acceptable for them to ride (Holt 2004). Females who ride may be more likely to be attracted because they seek the enjoyment that riding can give them on a more extreme level.

Features of Risk

There are three features that relate to risk; road features, distraction and other traffic. All of these features are, or are believed by the rider to be, external to the rider. These are third-party features. Even losing concentration can to be blamed on some third-party element causing a distraction.

Risk and Road Features

Road features have a high correlation to risk ($n = 291$; $r = +0.303$); as the road surface quality reduces perceived risk increases. With the unstable nature of motorcycles some of this increased feeling of risk may be justified (Institute of Highway Incorporated Engineers 2005).

> Hidden junction on bend ahead? – Scenario two. (Broughton 2007)

> Surface appears poor. Possibly traffic emerging from side roads and farm tracks.
> – Scenario one. (Broughton 2007)

Risk and Likelihood of Distraction

The likelihood of a rider being distracted has a positive relationship with risk. If riders feel that they are riding where getting distracted is a possibility, then their perceived level of risk is higher. Riders identified situations like watching out for speed enforcement cameras and busy urban areas, such as shown in scenario five, as being possible causes of distraction. One rider commented about scenario five that:

> Slow, busy, stop and go, with lots of distractions. (Broughton 2007)

Risk and Other Traffic

Other traffic has the highest loading on the risk task, around three-quarters of riders state that the risk to them would be high when they are riding in areas with a high density of other traffic (Broughton 2007). Figure 7.11 illustrates the strong relationship between overall risk and the risk from other traffic. When the author asked riders how often crashes were the fault of other vehicles over half believed that other vehicles causes 50 per cent or more of PTW fatal crashes (Table 7.6). Yet in reality only around 25 per cent of motorcycle fatal crashes are the fault of other road users (DfT 2006a).

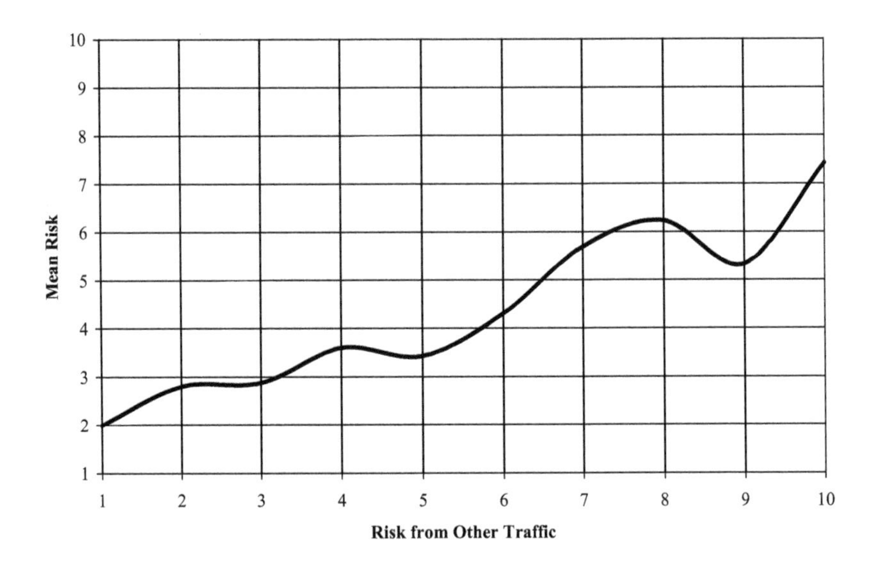

Figure 7.11 Risk from other traffic and mean risk

The belief that other road users are threatening to motorcycle riders is further emphasised by the following rider quotes:

> Lack of consideration from other road users. (Broughton 2007)
>
> Other road users who are very ignorant and sole purpose seems to be to cause accidents. (Broughton 2007: p. 294)
>
> Careless inobservant drivers. (Broughton 2007)
>
> Car drivers being inconsiderate and unaware. (Broughton 2007)
>
> Car drivers, always have to have your wits about you because of the car drivers. (Broughton 2007)
>
> Car drivers that don't look at junctions. (Broughton 2007)

External Risk

In examining the risk factor designated as 'external risk', it comprised road features (0.82), rider distraction (0.88) and other traffic (0.87), with risk having a weighting of 0.72. These elements are 'third party' to the rider and may be viewed as out of the rider's control with the exception of rider distraction. It could be argued that if the motorcycle rider was focused on riding then they would not be distracted. However, this element has scored high in the external risk factor, therefore some riders must consider that they do get distracted and this leads to higher levels of perceived risk. As the three elements are external to the rider then this factor is designated as 'external risk'.

Table 7.6 Fatal crashes involving bikes are the fault of other road users

Answer	%	Cumulative %
10% of the time	13%	13%
25% of the time	35%	48%
50% of the time	20%	68%
75% of the time	29%	97%
90% of the time	3%	100%
Total	100%	

Task Difficulty

As explained in Chapter 3, task difficulty has an important relationship with road user behaviour. A comparison of levels of task difficulty with the times that enjoyment is experienced, or when a feeling of risk is sensed, can therefore give insight into aspects of rider behaviour.

'Rush Based Enjoyment' tends to be found at lower task difficulties; as only a low level of skill is required to ride fast in a straight line this enjoyment can be found at low task difficulty levels.

'Challenge Based Enjoyment' has a different relationship to task difficulty than 'Rush Based Enjoyment' with high enjoyment being found at medium task difficulty and very low enjoyment at low and high task difficulty. This is a flow type enjoyment profile with maximum enjoyment being found at a level when skills are matched to the task difficulty. When task difficulty is low then an apathetic state is produced that is not enjoyable. At the other end of the scale when a high task difficulty event approaches the limits of the rider's skills, then anxiety results and this is felt as the non-enjoyable state of risk. This links with the expectation if riding is seen as a flow activity as described by Csikszentmihalyi (1990) (see Chapter 6).

Figure 7.12 reflects this flow interpretation. Low enjoyment from challenge is present at low task difficulty where the skill set is not being tested. At high task difficulty, where the skill set is being challenged, enjoyment also reduces. In the mid-range difficulties, where challenge is matched by the rider's skills, enjoyment is greater.

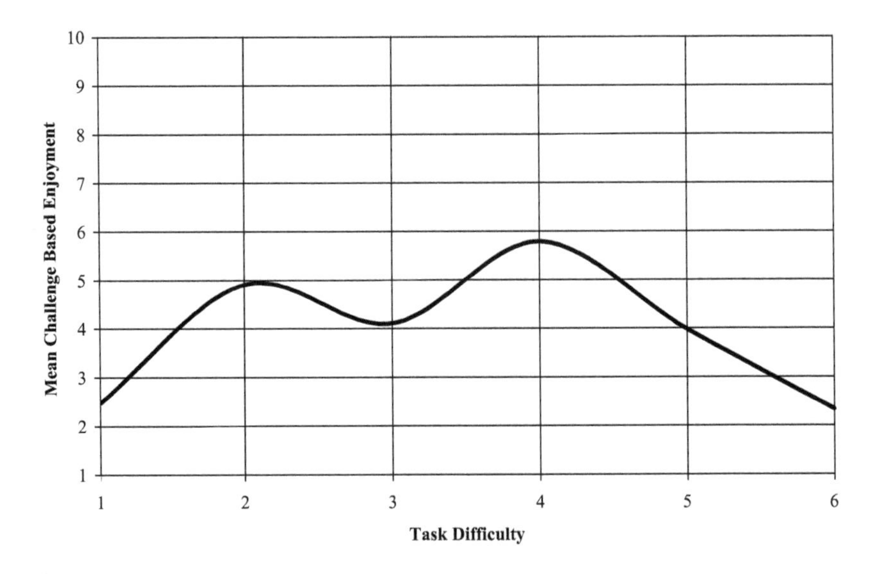

Figure 7.12 Mean challenge based enjoyment and task difficulty

Generally riders perceive more risk in areas of high task difficulty, with risk ratings being consistently lower with low and medium task difficulties. This is emphasised in Figure 7.13 where the increases in risk does not tend towards being linear. Rather risk swings between 3.5 and 4.5 for lower task difficulties, before rising steeply at a task difficulty of 5 and remaining high. This suggests that below a task difficulty threshold risk is fairly constant, but once that threshold has been exceeded then risk increases with a step function. This step function with the data is consistent with the theory of task difficulty homeostasis (Fuller 2005) where risk is seen as being low while the demand of the task is below the perceived level of the skills being used to undertake that task. However, when the task demand level exceeds the capability level then a person is in an out-of-control situation, with this being felt as risk. This was discussed more fully in Chapter 3.

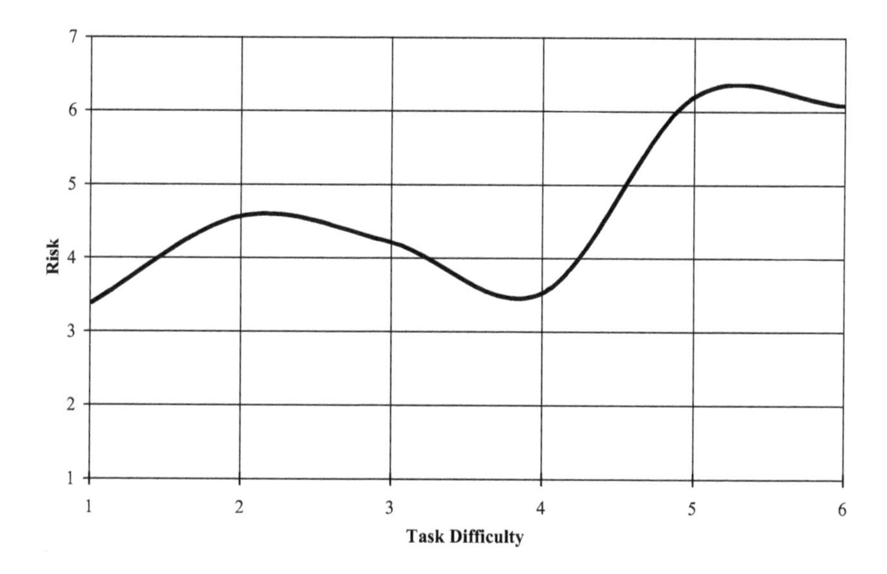

Figure 7.13 Mean challenge based risk and task difficulty

Summary

Enjoyment is found in one of two ways: 'Challenge Based Enjoyment' where bends are highly important; and 'Rush Based Enjoyment' with straight line speed as a major factor. 'Rush Based Enjoyment' is rated higher when riding is less difficult More riders tend to get enjoyment from 'Challenge Based Enjoyment' than from 'Rush Based Enjoyment', but younger riders are more likely to prefer 'Rush Based Enjoyment'.

Many factors are involved in determining rider risk and enjoyment. There is a relationship between challenge and enjoyment with lack of challenge most often being an enjoyment inhibitor. Speed is important for enjoyment but this relationship is a step function rather than linear, suggesting that a minimum speed, or task difficulty, is needed for enjoyment. Thus, for most riders, challenge rather than risk is the dictator of enjoyment but in seeking challenge riders may be increasing their risk. Up to now, this book has focused on the PTW rider to examine their attitudes, behaviours and motivations. While it has already been established that riding is a comparatively risky activity over driving, the question arises, is vulnerability to hazards the only difference? The following chapter seeks to explore this question by comparing motorcycle riders with car drivers in relation to how risk and enjoyment interrelate.

Chapter 8
Bike Riders and Car Drivers

The examination of how enjoyment and risk interact for motorcycle riders indicated that in situations considered risky by riders enjoyment was seldom found. Enjoyment was usually the product of the rider being challenged by the riding activity. This is reminiscent of what Csikszentmihalyi described as the flow state (Csikszentmihalyi and Csikszentmihalyi 1988). Enjoyment appears be found in two ways; 'Rush Based Enjoyment' (RBE) where speed-generated adrenalin rush gives rise to enjoyment; and 'Challenge Based Enjoyment' (CBE) where pleasure is derived through feeling 'at one' with the bike while meeting the challenges of the road. This second type of enjoyment is more closely associated with the flow type experience. Up to now the discussion has centred around the experience of Powered Two-Wheeler (PTW) riders with the argument that as most riding is expressive, then it is fundamentally different from most driving. The argument being that the purpose of most transport is to travel from A to B safely. While car driving may be quite enjoyable at times, it is mainly concerned with arriving at the destination for some other purpose. Although PTWs can serve this purpose, often riding is mainly about the journey itself. Therefore, when riding a PTW, the journey is more likely to be the primary fixation and the destination is of secondary importance. This contrasts to most driving where reaching the destination is the primary consideration and the drive is the 'means to an end'. If this is the case then it would be expected that the relationship of risk and enjoyment would be different. This chapter compares these aspects of riding PTWs with the experiences of driving cars to identify the similarities and differences between the two modes of transport.

Risk and Enjoyment

The view of some non-riders that PTW riders must gain enjoyment from the feeling of risk may be due to the vulnerable nature of motorcycling. Such views are demonstrated by the quotes below:

> People who like the thrill of adrenalin. Risk takers. Confident. Good drivers but prepared to take chances as more chance of injury on a motorbike than in a car. (Broughton 2007)

> Put enjoyment over personal risk, also inconsiderate of impact of their serious injury or death on family members. Usually associate it with men going through a middle-age crisis (like my husband!). (Broughton 2007)

However, despite this view the majority of motorcyclists do not ride because they enjoy the feeling of risk that PTW use can generate, rather they ride because they gain enjoyment in other ways, mostly through meeting the challenge that motorcycle use can generate. For motorcyclists, the relationship between risk and enjoyment is not a positive linear relationship; an increase in risk does not lead to a corresponding increase in enjoyment. The risk enjoyment profile for riders shows that enjoyment peaks at the mid-range of risk and for most riders little enjoyment is found at low levels of risk, but nor was enjoyment found at high levels of risk.

In order to test if this was the same for car drivers, Broughton (2007) ran a similar data gathering exercise with a group of car drivers. As with the PTW riders, the drivers were asked to rate the same six scenarios, described in Chapter 7, for risk and enjoyment. They were also asked their reasons for the ratings.

The corresponding driver profile showed a negative linear relationship. For the average driver high risk correlates with low enjoyment, and low risk with high enjoyment. Figure 8.1 shows a comparison of the rider and driver risk/enjoyment profiles demonstrating the differences between the two road user groups. Drivers' enjoyment decreases with an increase in risk, while rider enjoyment peaks at a mid-risk value. Generally drivers gain enjoyment in less risky situations compared to riders with the driver profile being similar to the rider 'Risk Averse' type, ergo the average driver tends towards being risk averse.

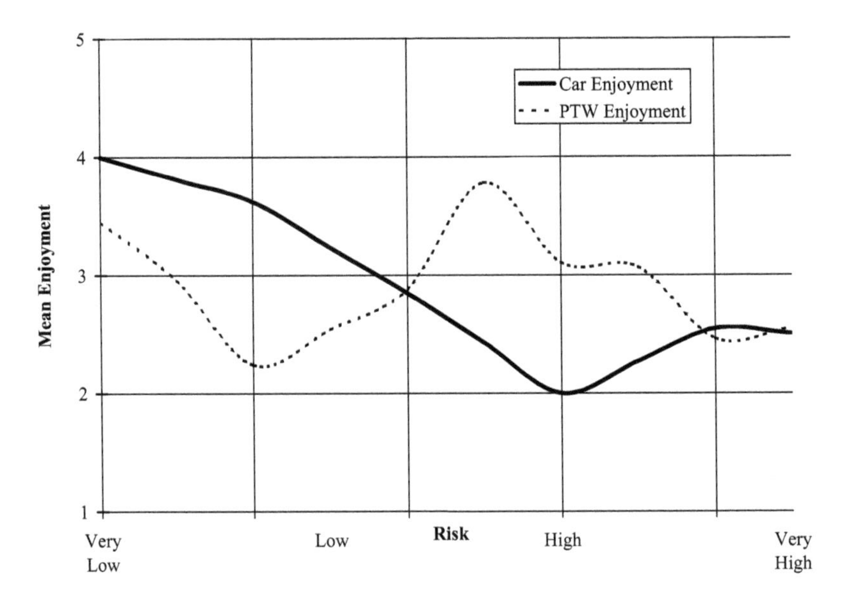

Figure 8.1 Risk and enjoyment for drivers and riders

Risk

There is a perceptible difference in how risk relates to the assessed factors for drivers and riders; analysis shows that there are only two common variables that have a high correlation – road features and other traffic.

A comparison of road features for the two vehicle types, with respect to risk, shows quite similar profiles with a linear increase in risk with respect to rated road features until a threshold is reached where risk plateaus at a constant level.

There is also a parallel relationship between car drivers and PTW riders when risk is compared to the ratings for other traffic (Figure 8.2). For both vehicle types risk increases in a linear fashion with respect to traffic density suggesting that both groups may feel that it is the actions of others that put them at risk rather than their own actions.

Three other elements correlated with driver risk: speed, temptation and visibility, all negatively. Drivers are therefore less likely to be tempted to drive enthusiastically, or willing to drive fast, in areas that are considered risky and they also feel that it is more risky when visibility is reduced.

In a comparison of the interaction between speed and risk for drivers and riders, it was found that when drivers feel that they are at low risk then they are more willing to drive faster. When driving at medium to very high speed, risk is constant suggesting that once risk reduces to a certain level then risk is no longer a speed inhibitor. For riders, the relationship between risk and speed is less clear as there is not a linear relationship.

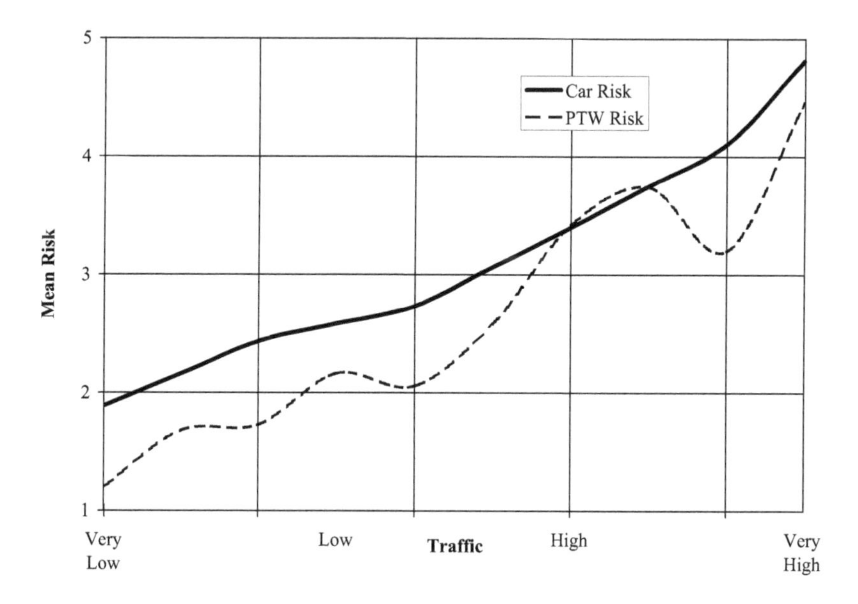

Figure 8.2 Risk and other traffic for drivers and riders

While this is clearly demonstrating differences in the enjoyment/risk relationship, further analysis by Broughton (2007) found that the elements that generate feelings of enjoyment and risk are also different for drivers and riders.

Enjoyment

Of the seven variables that correlate with enjoyment for car drivers, five also correlate for riders: Road surface quality; bends; temptation; speed; and surroundings. Other traffic and road features correlated with driver enjoyment but not for riders. However, visibility and challenge correlate for riders but not drivers.

Both drivers and riders find more enjoyment in areas that have a better road surface quality. Similar profiles also exist with respect to bends when comparing drivers and riders (Figure 8.3), despite the challenge variable not correlating with enjoyment for drivers. The lack of correlation gives the impression that drivers may not believe that driving around bends is challenging their skill set in a rewarding way.

One of the elements that relate to both risk and enjoyment for drivers is speed, with a positive correlation for enjoyment and a negative correlation for risk. Figure 8.4 compares enjoyment for drivers and riders in relation to speed. For car drivers, enjoyment increases linearly. For riders, enjoyment is a step function occurring at a mid-speed rating.

As with PTW riders, the data on car drivers was analysed in more depth through the data reduction technique of factor analysis (see Chapter 7 for more details).

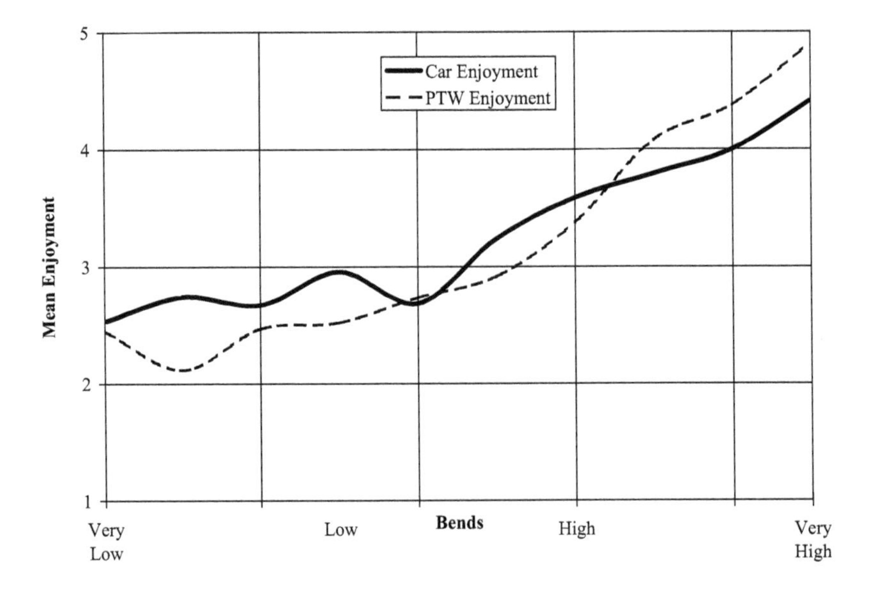

Figure 8.3 Enjoyment and bends for drivers and riders

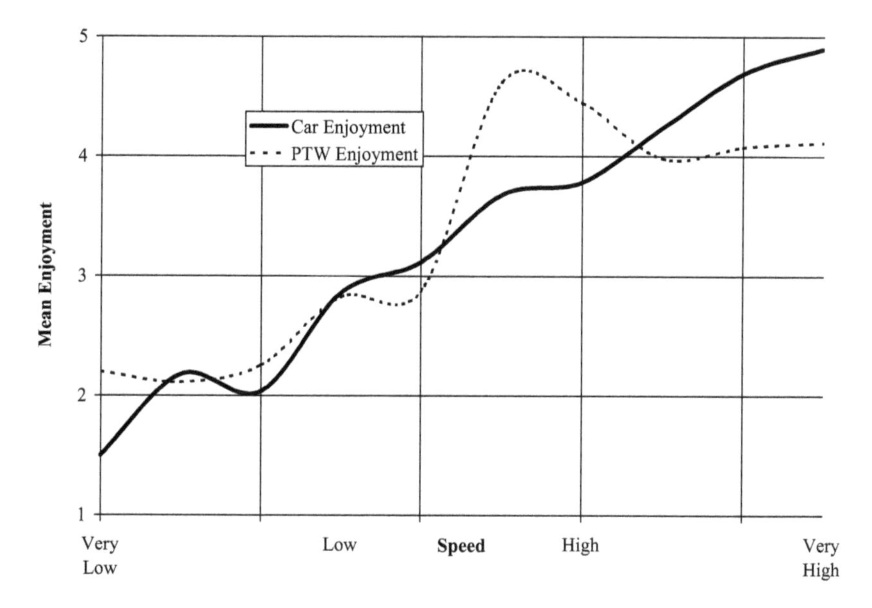

Figure 8.4 Enjoyment and speed for drivers and riders

Enjoyment and Risk Types

In a similar way to PTW riders, car drivers split into three factors, two relating to enjoyment and one to risk. The driver RBE contains the same elements as the equivalent PTW factors with the exception of road surface quality. This was not included for car drivers almost certainly because it has considerably less impact on vehicle handling and safety for car drivers than PTW riders. The enjoyment weighting for car drivers in the RBE factor is somewhat higher than its corresponding PTW factor (0.60 and 0.48 respectively) implying that RBE is a more significant way of gaining enjoyment for drivers.

Driver CBE is also very similar to the equivalent motorcycle factor, but without the temptation element. Driver 'external risk' is also similar to rider external risk except that a negative correlation to surroundings is included.

The inclusion of road surface quality for riders in RBE shows that riders are less likely to ride fast unless the road surface quality allows it. Drivers may feel that cars are stable enough to drive fast in a straight line, even with poor road surface quality.

Males tend to find driving more enjoyable than females, however, there is no discernible difference between the genders for motorcycle riders. Female riders are more likely to gain RBE than their riding male counterparts but male drivers are more likely to find RBE than female drivers. As discussed earlier, this may in part be explained by the type of females attracted to riding and its association with activity choice while car driving is more likely to be a functional means of getting from A to B and not seen as an activity of choice but as a necessary element of achieving other goals such as travelling to work or accessing leisure opportunities.

Target Task Difficulty

One of the key elements in Fuller's (2005) model of Task Homeostasis is that drivers select a target task difficulty which they aim to achieve. This was posited to be the same for riders by Broughton (2008a). How close this target level is to their perceived capability defines the safety margin that they are willing to accept at that time. One of the differences between drivers and riders affecting this target task difficulty is trip purpose; most PTW use is expressive, that is riders go out purely for the enjoyment of the ride (Broughton 2005; Hannigan, Fuller, Bates, Gormley, Stradling, Broughton, Kinnear and O'Dolan 2008). Therefore, as speed is linked to enjoyment for riders (Broughton 2008a), they tend to enjoy riding faster in comparison to drivers (Hannigan et al. 2008). As riders are more likely to be going 'A to A' rather than 'A to B', there is less likely to be destination-generated time pressures. This suggests that there is a high probability that the reason for this higher speed is linked to pleasure rather than logistics. Riders tend to set a higher target task difficulty level when 'free' riding in a rural environment and, therefore, will be riding with a lower safety margin than their driving counterparts.

Within the urban environment riders tend to lower their target task difficulty; this may partly be due to an increase in perceived risk due to the hazards presented by other vehicles. These potential third-party hazards would raise the rider's own assessment of task demand due to a requirement of a higher level of hazard perception (Broughton and Stradling 2005).

Riders are more likely to respect urban speed limits and exceed limits on the open road with this speed being related to a higher target task difficulty. This higher level of target task difficulty within a rural situation is in part set by the goals of the recreational rider – enjoyment. The lower target task difficulty in an urban environment is partly due to perceived risks caused by third parties. The perceived increase in task difficulty is to a degree caused by the required higher level of hazard perception. Riders adjust their target task difficulty to compensate for what they see as threats. These threats are almost all third party (Broughton 2006) and therefore when riding is taking place in an area with low third-party risk, such as the open road, the margin of rider safety is smaller because of the high target task difficulty.

Young Drivers and Riders

Young drivers and riders are over-represented in Killed and Seriously Injured (KSI) crashes (DfT 2004c; DfT 2005b; DfT 2006a; Harre 2000; Stradling 2005). This higher risk for young people warrants investigation to uncover any differences between the young and old, but also between young motorcycle riders and young car drivers. The similarities between young drivers and riders may also reveal insight into attitudes and behaviours. This may reveal patterns to assist in understanding both groups and aid in the design of interventions.

The data collected by Broughton (2007) suggests that the level of enjoyment is different for younger (under 26 years old) road users. Young drivers are more likely to find driving highly enjoyable than older drivers. A substantial number of older drivers were found not to enjoy the driving experience at all, suggesting that their driving is more out of necessity than a desire to drive; need rather than want. While there is a reduction in PTW riding enjoyment with age, this is less marked. The level of driving enjoyment for young drivers is higher than the riding enjoyment claimed by young riders. This may be in part due to the higher skill level required to feel comfortable in riding. The level of restrictions faced by young riders which reduces the performance of bikes they can ride may also have an impact.

How though, is this enjoyment for young drivers and riders generated?

Young drivers are more likely to experience RBE compared to older drivers with a similar but less marked distinction seen for PTW users.

There is a plethora of publications that demonstrates that younger riders and drivers are more liable to be sensation seekers or risk takers (Clark and Ward 2002; Stradling 2005) with this type of behaviour pattern being reflected in attempting to gain RBE. Speed is one of the main factors related to RBE. Both young riders and young drivers tend to ride or drive faster than those who are older. Older riders tend only to ride slightly slower than younger riders while older drivers drive substantially slower than younger drivers. Younger riders and drivers do not want to drive or ride slowly, but young riders tend to be more conservative in their speed choice than young drivers. This speed choice difference is a reflection of the extra vulnerability felt while riding coupled with the added difficulty of the riding task.

Summary

There are appreciable differences between motorcycle riders' views of risk and enjoyment compared to those of car drivers. For drivers, as risk increases enjoyment decreases, while for motorcycle riders the relationship is more complex with a peak of enjoyment occurring at a mid-risk point. There is also a clear relationship between the speed a driver would be willing to drive at and risk, with drivers going slower as risk increases; this is not seen for riders.

However, many similarities exist for the two road user groups with both experiencing enjoyment in one of two ways, RBE and CBE. For RBE, drivers did not have the road surface quality element as a factor and the temptation element was not present in the drivers' CBE factor. Therefore, although the enjoyment types are similar, they consist of slightly different components.

Further exploration of the demographics of the two groups revealed a significant difference in enjoyment gender profiles; males enjoy driving more than females while there are no discernible gender differences for riders. This dissimilarity between riders and drivers may be due to females who do not enjoy driving, but feeling that they have to drive; it is a functional activity based on other life choices or commitments. Riding is more of a choice activity and females who ride do so because they want to. Female riders are also more likely to want to experience RBE than males, with the opposite being true for drivers. This may also be an effect of riding being a choice activity and females who choose to ride may be attracted to riding because they are sensation or thrill seekers.

As a general statement, riding is more enjoyable than driving, however, younger drivers experience higher levels of enjoyment than young riders do. This chapter shows that there are some similarities between riders and drivers, but there are also appreciable differences, such as how risk and enjoyment interact. Such differences will be of particular importance when considering ways of improving road safety for different road users. The following chapter revisits Fuller's model of Task Difficulty Homeostasis in the light of the risk and enjoyment relationship.

Chapter 9

Task Capability, Task Demand and Motorcycle Riding

Introduction

Having explored issues of risk and enjoyment, and the way that they interact for riders and drivers, this chapter returns to issues linked to task difficulty. Fuller's model of Task Difficulty Homeostasis was discussed in Chapter 3. As described earlier, task difficulty is an important aspect of road user behaviour and therefore merits a discussion on how motorcycle riding is affected by it.

According to Fuller (2005):

> Task difficulty arises out of the dynamic interface between the demands of the driving task and the capability of the driver. (Fuller 2005: p. 463)

In Chapter 3 the components that compose the task of driving were identified and revised to account for the differences of the riding task. The 11 component tasks are listed and described in Table 9.1. The sum of the demands for these individual tasks gives the total task demand.

If a rider's capability exceeds the demands of all the tasks being undertaken during riding then the rider will be in control. If the capability is lower that the total task demand then loss of control results. Figure 9.1 illustrates this loss of control resulting when Capacity is less than Demand (C<D), culminating in either a lucky escape or a collision.

The results of Fuller's (2005) task difficulty experiment are plotted in Figure 9.2 where a scenario was used to assess driver speed, task difficulty and risk.

For a given scenario, task difficulty and experience of risk is related to speed. Statistical risk (risk of having a crash) is zero until a speed threshold is reached, and then rises in a linear fashion.

Figure 9.3 is the same data that is shown in Figure 9.2, but plotted against task difficulty rather statistical risk. This indicates that the risk experienced has a linear relationship with task difficulty. Estimated crash risk is zero until a task difficulty threshold of just over 4 is reached, then this type of risk rises proportionally to task difficulty.

Table 9.1 Eleven components of the riding task (Broughton 2007)

Task	Description
Strategic levels	Activity choice (functional and/or expressive) Departure time choice, route alternatives and travel time
Navigation tasks	Find and follow chosen or changed route
Hazard perception	Detection of hazards
Road tasks	Choose and keep correct position on road: road position may be modified by road surface quality hazards
Traffic tasks	Maintain mobility ('making progress') while avoiding collisions (reaction to hazards)
Rule tasks	Obey rules, regulations, signs and signals
Handling tasks	Use PTW controls correctly and appropriately Interaction of PTW and rider (leaning at corners, etc.)
Secondary tasks	Keeping visor clean/demisted; acknowledgment of other riders; using satellite navigation equipment
Speed task	Maintain a speed appropriate to the conditions; speed will be modified by hazard perception
Mood management task	Maintain driver subjective well-being, avoiding boredom and anxiety
Capability maintenance task	Avoid compromising driver capability with alcohol or other drugs, fatigue or distraction

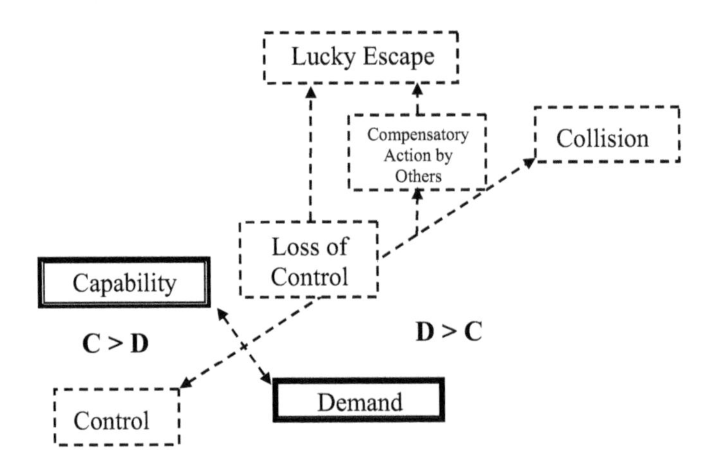

Figure 9.1 Outcomes of the dynamic interface between task demand and capability (Fuller 2005: 464)

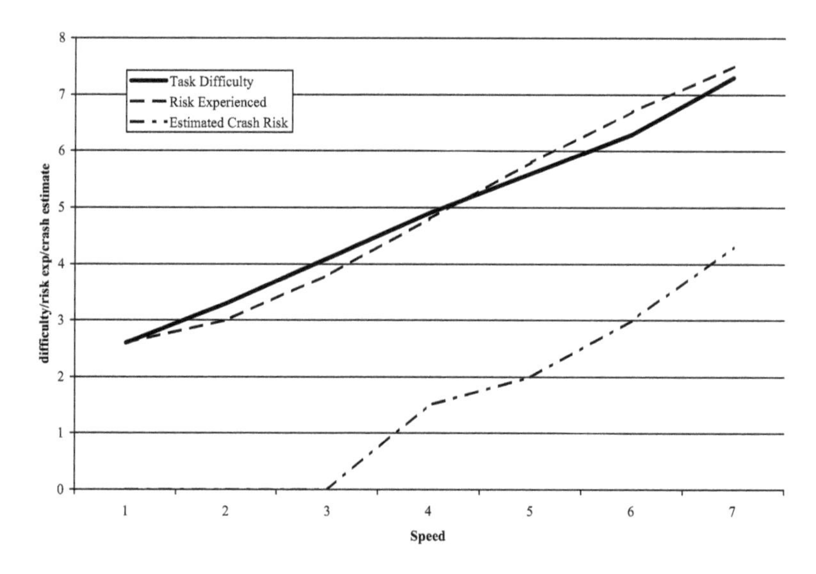

Figure 9.2 Ratings of task difficulty, estimates of crash frequency and ratings of risk experience (data extracted from Fuller 2005: 469)

Task Difficulty and Riding Enjoyment and Risk

Broughton (2007) carried out similar data collection to Fuller (2005) except the data was collected from motorcyclists in various scenarios. The Risk and Enjoyment profiles from Broughton's data, with respect to task difficulty, are shown in Figure 9.4. This graph demonstrates a similar profile to that noted by Fuller with risk being low (below 3) until point A where a threshold is reached that causes a large rise in risk; at a similar point there is a large curtailment in enjoyment. This demonstrates that the theory of Task Homeostasis developed for car drivers (see Chapter 3) is also applicable to Powered Two-Wheeler (PTW) riders.

For car drivers, the implication of this is that additional safety interventions are required that assist in reducing crashes. Training, safety features in cars and awareness raising of risk factors can be used to lower incidents of Killed or Seriously Injured (KSI) crashes for car drivers. However, while Task Difficulty Homeostasis Theory may hold true for PTW riders, the implications for interventions may be different. As PTW riding and car driving have been shown to be fundamentally different in nature for most riders, the issue of enjoyment and its implications needs further exploration. As discussed earlier, for most PTW riders, journeys are often expressive, even if functional elements are involved, while most car journeys are predominately functional and are undertaken for a purpose other than simply to drive. For riders the use of PTWs as a mode of transport is an active choice that makes it more closely related to a leisure activity than a utilitarian activity. This being the case, the enjoyment element of riding may be more important. For drivers

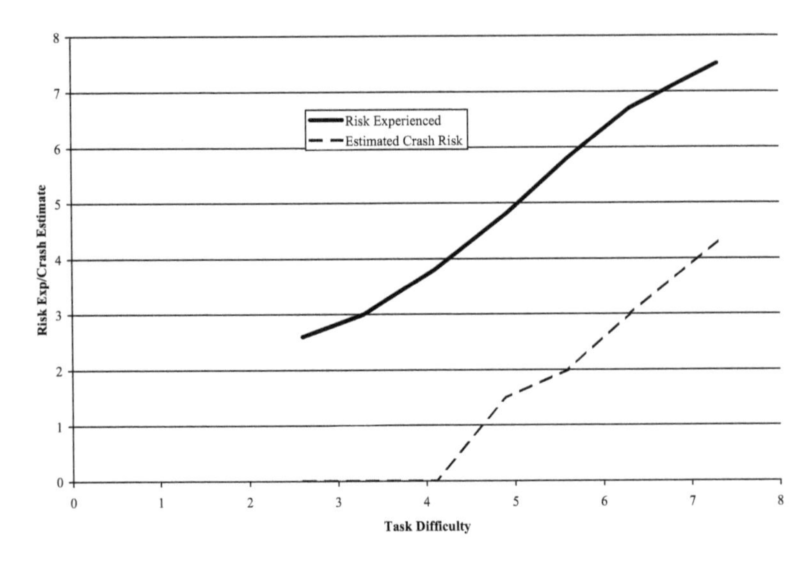

Figure 9.3 Estimates of crash frequency and ratings of risk experience with ratings of task difficulty (data extracted from Fuller 2005: 469)

a reduction in task demand is generally seen as a good thing, as it helps them go from A to B in a safer manner. While safety is a real issue for PTW riders and risk is not actively sought by most riders, there is a need for challenge to maintain the enjoyment experienced in PTW use. In Chapter 6, Csikszentmihalyi's Theory of Flow was introduced; this was used to explain aspects of PTW use. It may also assist in understanding riders and the type of interventions that may help in reducing KSI crashes without taking away the things that makes the activity enjoyable.

Task Difficulty and Flow

Csikszentmihalyi's (1990) Theory of Flow describes four states:

1. Apathy – resulting when both skills and challenge are low.
2. Boredom – resulting when skills are higher than the required challenge.
3. Anxiety – resulting when skills are lower than the required challenge.
4. Flow – the state entered into when skills and challenge are matched.

As riders have a reasonable skill level due to the level of training required prior to riding on public roads it is unlikely that the low state of skills and challenge required for an apathetic state will exist. As challenge is being compared to skills, high difficulty of a task could also be described as challenging. The flow state itself is highly enjoyable with boredom and anxiety being less so, therefore if skill level is assumed to be constant then enjoyment can be plotted against difficulty. Note that for riders the state of anxiety entered into when skills are outstripped by challenge would be felt as risk.

The profile shown in Figure 9.4, where risk increases at the same point that enjoyment decreases, can be expressed in a model with respect to task difficulty (Figure 9.5).

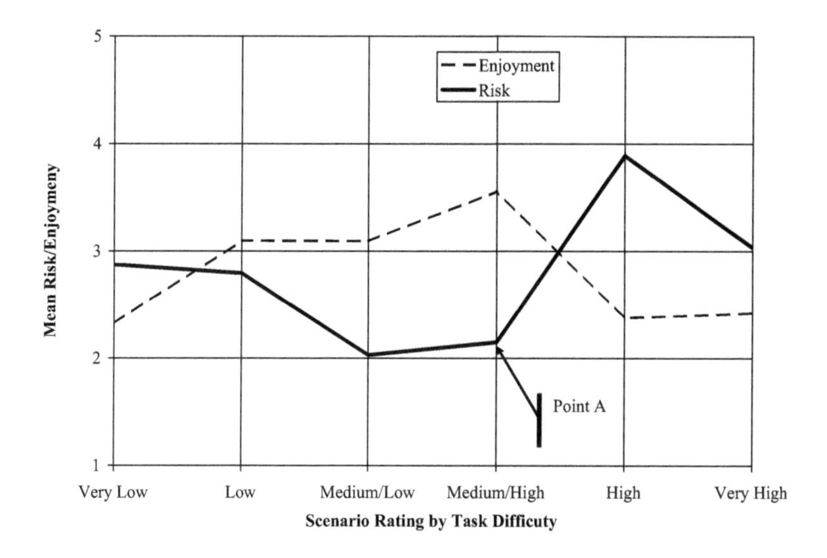

Figure 9.4 Risk and enjoyment by task difficulty (all scenarios)

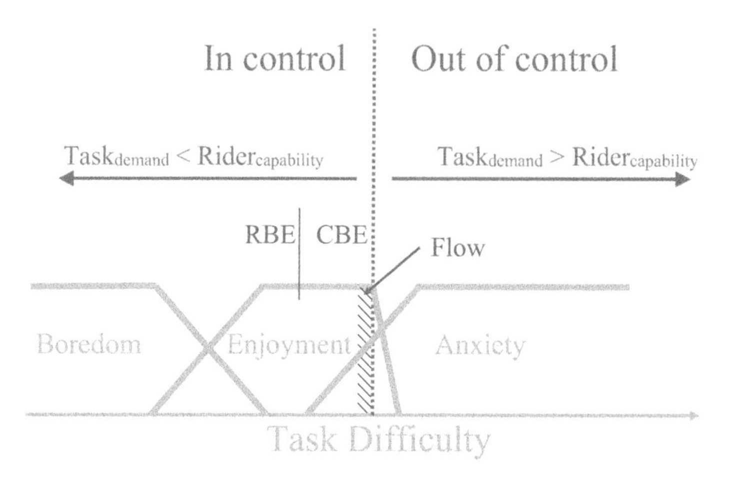

Figure 9.5 Linear relationship of flow states and task difficulty (Broughton 2007, 2008)

This flow model shows that task difficulty is not only related to risk, but also to enjoyment. Once a threshold of task difficulty is reached then enjoyment declines and risk increases. As task difficulty is a function of task demand and rider capability then a change in either of these will have an effect on task difficulty. The task demand element can be affected directly by the nature of the road being ridden. How though do these elements interact with task difficulty?

Task Difficulty and Road Elements

The 11 elements previously identified as being related to risk and enjoyment can be split into two groups, those related to the external environment:

1. road surface quality;
2. road features;
3. visibility;
4. other traffic;
5. surroundings;
6. challenge;
7. bends;
8. likelihood of distraction.

And those related to internal rider state factors:

9. temptation to ride enthusiastically;
10. speed;
11. overtaking.

The model of task difficulty shows that task difficulty is moderated by controlling task demand and capability. The main way for a rider to increase or decrease task difficulty is by altering their riding speed, the faster they ride the higher the task demand and therefore the higher the task difficulty. Task difficulty can be increased by escalating rider capability and one way this can be done in the long term is by training the rider. However, rider capability can be altered in the short term if the rider uses the more efficient implicit memory.

Implicit and Explicit Memory: Interaction with Capability

The earlier discussion on how task difficulty and flow interact has the state of flow being entered just before task demand begins to approach the limits of rider capability. In the discussion of neuro-cognitive mechanisms that underpin flow experience, Dietrich (2004) explains that for flow to exist then the activity being undertaken must be run exclusively using implicit memory. However, Horswill and McKenna (2004) suggest that some conscious effort, or explicit memory use, is required for hazard perception. This implies that a rider who is in flow cannot be employing hazard perception techniques, or at best some very rudimentary heuristic version that can be implemented implicitly. Also, when a rider is in a near flow state then only a small amount of the explicit memory is available to carry out the hazard perception task.

When a rider is operating mostly on 'automatic', that is, using implicit memory to achieve tasks, then the decisions made concerning potential threats have to be made quickly by the 'hazard stimulus' triggering a schema that implements a course of action; for example, braking when one sees the brake lights of the vehicle in front activate. When the stimulus does not align with one of these simple, but well-practised, schemas then a higher level of cognitive demand is required to decide on what action is required (Klein 1998). The level of automation will have to decrease to allow for this; therefore rider capability will also decrease. This sudden reduction in capability means that task demand may exceed this lower capability level and create an 'out-of-control' situation (Figure 9.1). The resulting out-of-control state can culminate either in a lucky escape or a collision; therefore the main causes of PTW crashes could be explainable by task demand exceeding rider capability.

Task Difficulty and Crashes

This section examines two types of common PTW crashes identified in crash statistics; loss of control on bends and crashes while overtaking, and applies Task Difficulty Homeostasis to explain them.

Loss of control by the rider on bends is a major cause of KSI crashes. Clarke, Ward, Bartle and Truman (2004) reported that loss of control crashes on bends accounted for around 12 per cent of all crashes, 7 per cent on left hand bends and 5 per cent on right hand bends. In a similar study that looked at Scottish PTW crashes between 1992 and 2002, Sexton, Fletcher and Hamilton (2004) reported that 9 per cent were 'going ahead on a right hand bend' and 11 per cent 'going ahead on a left hand bend'. Bends are a major factor in 'Challenge Based Enjoyment' (CBE). Riders may actively seek roads that have potential to provide such experiences. As this type of enjoyment is flow based, riders will be attempting to match their skill level with the demands presented by the environment, in this case, the bends. However, if the rider fails to assess either skill level or task demand accurately, or if an event occurs that suddenly increases task demand, or reduces rider capability, then loss of control will result.

Another common PTW manoeuvre being carried out during a crash is overtaking. Sexton, Fletcher and Hamilton (2004) reported 9 per cent of PTWs were carrying out this manoeuvre just prior to the crash and Clarke, Ward, Bartle and Truman (2004) reported a figure of 14 per cent in their study. Speed is an enhancer of task difficulty, and also a major element of 'Rush Based Enjoyment' (RBE). As PTWs can generally accelerate significantly quicker than cars, task difficulty can rapidly increase to a point where task demand exceeds rider capability. The resulting loss of control can occur before the rider is aware of what is happening or has time to reduce task demand by actions such as reducing their speed. These two examples show that a sudden change in capability or difficulty can place a rider in an out-of-control situation.

Summary

Developing an understanding of the different ways that enjoyment is obtained and risk perceived is an important step in understanding the goals of PTW users. When this understanding is coupled with an appreciation of task difficulty, then the information can be used to assess possible reasons for PTW crashes. This deeper understanding of why crashes may occur rather than simply how they occurred may assist in the development of more effective interventions.

Task difficulty is the interaction of task demand with rider capability, and when a rider matches task demand with their capability then the flow state can be entered into. Being in the state of flow implies that the rider is operating in a fully automatic mode, with this mode leaving little cognitive ability for other tasks such as hazard perception. Therefore if a sudden need to react to an unexpected hazard occurs then task demand is likely to increase rapidly and may exceed capability. This may occur when enjoyment is sought through challenge (CBE) and the rider over-estimates their capability or under-estimates the challenge faced.

For those experiencing RBE, speed is a key element, but one that raises task demand. The increase in task demand through excessive speed may result in task demand exceeding the rider's capability, resulting in loss of control. This helps us to understand individual rider behaviours and may assist in developing interventions targeted at PTW users. However, as discussed in Chapter 5, despite PTW riding mainly being individual, that is, usually only one person rides a PTW, riders when describing their feelings about PTW riding often stated ideas relating to community, sense of belonging and social aspects of riding. This feeling of being 'part of something' is often expressed through group riding. Riding as part of a group or informally with friends is a common element of the PTW leisure experience and one that may have different characteristics to the lone experience. The next chapter examines some of the psychology behind group riding.

Chapter 10
Group Riding

Social Groups

Many riders feel that they belong to the greater collective of motorcyclists. When out and about this sense of belonging is often demonstrated by acknowledging other riders with a wave, nod or flash of the headlights. It is also demonstrated in the number of forums, clubs and events for motorcyclists discussed in Chapter 5. This social aspect is taken further by some riders by riding as part of a group. This can be seen as an extension of the rider belonging to a collective and having a sense of belonging. Rider identity can be amplified by this sense of belonging such that the rider is not just another road user but a motorcyclist who is using the road. When riding specifically with a group, the rider is actually operating in the social group of the immediate riders but also within the greater social group of all motorcyclists. This group riding in some ways is a paradox as the actual riding of a bike is generally a solo activity with limited opportunity for communicating. When a rider is within a group environment, the dynamics of riding may change. The rider is no longer a solo road user making choices but rather the rider becomes a part of the collective of the group and this may affect the riding choices that are made.

Group Behaviour

Group membership gives people social identity. It can define who we are. When a person is associated with a group, they categorise themselves as a member. A member of a group will probably start to take on the attitudes and norms that they believe are part of the group (Turner 1991). Therefore with a riding group, those who join may take on the attributes that they believe the group holds. As those riding in a group are a sub-group of motorcycle riders they may take identity from the believed norms of motorcyclists in general as well as of the believed norms of the group they are riding with.

Everyone hold beliefs. Each motorcyclist will have a set of beliefs concerning how they should ride. Beliefs that are held need to be verified by the belief holder and this is done by using others to reinforce these beliefs (Festinger 1950). This need to have beliefs verified is one of the drivers of uniformity within a group. Another driver of group conformity is that group members want to avoid ridicule from other members (Deutsch and Gerard 1955). Therefore if a person's beliefs are not being verified by the group they will feel that they should adopt the beliefs of the group. If a rider becomes a member of a group, where the members ride in a more risky manner than the rider may desire, then the rider has to weigh up the risk of being ridiculed by the group if the group norms are not conformed to, with the increased risk of physical injury that may occur if the norms are complied with (Broughton 2007).

The pressure within a group can force its members to conform to group norms. A rider's intention to ride in a certain manner may be unsuccessful as the pressure within the group could adjust rider behaviour. However, the process that brings pressure to bear on a group member to conform to the group norm is not straightforward; it is complicated by the status relationships within the group and the personalities of the group members (Brown 2000). However, one way that this pressure may be brought to bear is by members going along with the beliefs or behaviours of the group and ignoring their own judgement in the process (Asch 1956). This desire to conform to the group norm may be so strong that some may conform in activities they know are wrong in order not to be different from the rest of the group (Festinger 1953). A rider who is riding with a group may have to change their behaviour from their normal riding style so that they comply with the norms they believe the group has.

With a formalised riding group, such as a club, there is less likely to be ambiguity as to the expected norms of the group as these are often set down in the club rules and regulations. This clarity means that riders are less likely to feel that they have to conform to less safe beliefs that others in the group hold (Brown 2000). In contrast, informal groups, like a few mates meeting up for a ride, will have no fixed rules and therefore ambiguity regarding the group norms may exist. With no fixed norms the group will take its riding norms from the beliefs of the group, with pressure for riders to conform to this. The norm itself will be shaped by the beliefs of the norm motorcycle behaviour; the beliefs from this will be formed, in part, by the image of riders portrayed in the press as 'the average rider'. With the media, including the motorcycle press, sometimes portraying motorcyclists as high speed riders then this image may be a major influence on the norms formed by an informal group and hence the individual's behaviour while in the group.

Audience and Coaction Effects

Within sport it has been recognised that the effect of an audience watching a performance is to increase the arousal level of the performer (Cottrell 1968), which in turn may increase competitiveness (Gill 1986). Arousal, which is the general state of activation, can range from deep sleep to being hyper with excitement. The upper ends of arousal are often being felt as anxiety (Gill 1986). Thus, arousal of a person may have a positive, or negative, effect on performance and an audience can alter the level of arousal.

The situation is further changed when the audience is comprised of those who are taking part in the activity with you (coaction). Within this situation, the performer may feel that they are being judged, not only be their peers, but by experts. Cottrell (1968) expressed the view that it is the presence of an expert who the performer believes can evaluate their performance that can raise arousal. The effect that arousal has on performance is more likely to be detrimental when the task being undertaken is difficult and complex (Gill 1986).

Although audience and coaction effects are mainly used to explain sports performances, these aspects may be of interest to those who are studying motorcycle rider behaviour, especially group riding. When riding in a group, it is likely that most riders would believe that at least some of their cohort would be in a position to make expert judgements on their riding; therefore this may well give rise to heightened levels of arousal. If a rider is concentrating on the way that he feels he is being judged by his peers rather than the riding task in hand, then the riding performance will be altered to reflect this; that is, there will be a change of immediate riding goals with the goal of impressing riding peers becoming predominant.

Group Riding and Riding Goals

If a rider feels that they are being judged or put under some subliminal pressure by their peers then the riding goals may change with the new immediate goal being to impresses the peer rider(s). This may involve displaying behaviours or skills that are believed to be required as part of the group.

If a person focuses the majority of their effort upon this immediate goal then a state of heightened anxiety may be generated (Easterbrook 1959), which in turn can negatively affect performance of skills (Wang 2002). Performance is affected as the performer narrows their attention to mainly focus on the new immediate goal. This is often accompanied by the narrowing of peripheral vision to allow concentration on the visual aspects that are felt to be important in achieving success as measured by this new goal. As performer attention is focused inward, upon themselves and not upon the environment in which the task is being undertaken, this may act as an inhibitor to the detection of, and the reacting to, information within the environment that is vital to good performance. For riders, not being fully focused upon the riding task, or not reacting to vital cues within the riding environment, can result in a mistake that could have serious ramifications.

A rider who wants to be respected by riding mates will have to decide what style of riding is required to create the correct image. This could lead to the use of the stereotypical image of 'risk takers' to base their behaviour on, and they may therefore ride in a more dangerous manner than before. Equally, it could be that riders decide that the best way to demonstrate their skill levels to their cohort is by riding fast, probably faster than they would ride if they were not under assumed peer pressure.

The need to 'prove' to others that they are riding with a high level of skill may be seen as riders competing with each other. Competition can have positive effects on performance, but can also have negative effects, for example, competition can improve the speed of the task in hand, but in doing so accuracy may be reduced (Church 1968).

Task Homeostasis and Group Riding

If this situation is considered using the Task Homeostasis Model (Fuller 2005) then there are two main issues: capability and demand. As discussed earlier, task demand is the sum of the demand of all the tasks being undertaken at one time. If a rider is focusing highly on a goal of 'looking good' to the rest of the group then this will be an extra task and due to the high importance of the task in the rider's mind, it will be given a lot of resources. This will reduce the amount of task resource that is available for other riding tasks. The rider may also be going faster to demonstrate that they are highly skilled; this increase in speed will also increase the task demand on the rider. In a situation where a rider is pursuing a goal motivated by peer pressure, they are very likely to find themselves in a position where task demand outstrips their capability and thus be in an out-of-control situation.

Group Riding Behavioural Change

In conversations with traffic police the first author was informed that some officers believed that riders towards the rear of the group often put themselves at a higher level of risk by carrying out dangerous manoeuvres to keep up with the group. In this situation, the immediate goals of the rider may be to keep up with the group as by not doing this the rider may lose the respect of the group by 'not being good enough'.

In an experiment that was conducted by the first author, riders were asked to rate how often they would overtake a vehicle in a specific scenario while riding by themselves or at the back of a group. The photograph scenario used is shown in Figure 10.1

The analysis of this data showed that there were three distinct categories of rider with respect to group riding: those who would overtake more often when group riding; those who would overtake less often; and those where overtaking behaviour would not be affected at all (Table 10.1). This shows that the effects of riding in a group can influence a rider to act in a more risky or a less risky manner, or have no affect at all.

Around 19 per cent of the riders in this study were more likely to take risks by overtaking because of the group influence and about 23 per cent being less likely to do so. However, those taking the extra risks involved with overtaking the car did not see it as being more risky, but rather felt that overtaking while riding in a group

Figure 10.1 Photograph used to assess group riding (courtesy of Owl Research Ltd)

Table 10.1 Percentage of riders by overtaking category

Category	**Per cent**
OT more in group	18.7%
No difference	58.8%
OT more by self	22.5%
Total	100.0%

was less risky than when riding alone. They also believed that the risk of being involved in a crash or near miss while performing the overtaking manoeuvre was also lower when they were riding in a group. These riders demonstrating riskier behaviour had a lower safety margin while riding and also declared that they liked to take risks. Those who are in the category (Table 10.2) of being more likely to overtake while in a group would do so faster with a mean speed of 66.0 mph compared to a mean speed of 60.9 mph for those who are more likely to overtake when riding alone – $t(173) = 2.219, p = 0.030$.

If a rider believes that they belong to the group that they are riding with, then they are less likely to be in the 'overtake more when riding alone' category. This is a reflection on accepting the group norms, where a group membership belief is

Table 10.2 Belonging to the group with overtaking behaviour ($\chi^2(df = 2; n = 170) = 9.15; p = 0.01$)

When riding in a group	'I feel that I belong to that group'		
	Not agree	Agree	Total
Overtake more often	9%	26%	19%
No difference in overtaking	59%	55%	57%
Overtake less often	32%	19%	24%
Total	100%	100%	100%

held. In this type of situation, fear of ridicule for not keeping up with the rest of the riding cohort may be a strong influence on this behaviour (Brown 2000). Those who were more willing to overtake when group riding also said that they felt that they had to fit into the crowd rather than be their own person.

How a rider feels about the others in the group also affects how safe the rider feels with those who feel that it is safer to ride in a group being more likely to feel that they have the respect of other riding group members ($n = 171; r = +0.286$). However, there is a dichotomy as those who feel that they are safer while riding in a group also take more risks when riding ($n = 169; p = +0.247$).

Summary

When motorcyclists choose to ride as part of a group, it may intensify the rider identity, as well as adding to the enjoyment of the ride activity. Group riding, although mostly carried out for social reasons, has safety implications due to the way a rider reacts to performing in the company of peers. Riders who feel that they have nothing to prove to these peers may take fewer risks when group riding than they would when riding alone. However, if a rider feels that he has to prove that he is part of the group, then riskier behaviour may be employed as a tool to accomplish this.

While group riding and its impact on rider behaviour is under researched, it clearly does have an effect. Whether riding as an individual or as part of a group, Powered Two-Wheeler (PTW) users are a high-risk group amongst road users. Understanding the dynamics of rider behaviour in lone or group situations may assist in designing more effective interventions based on the psychology of riding and the rider goals.

Chapter 11
Improving Motorcycle Safety

Introduction

As motorcycle riders are vulnerable road users, methods and interventions to reduce their casualty rate are always being sought. For an intervention to be effective an understanding of the nature and type of crash that they are exposed to, as well as any elements of rider behaviour that might place them at greater risk, is needed. While the statistical evidence on where, how and why crashes occur is useful in assessing the crash itself, methods designed to reduce the number and severity of such crashes require an understanding of rider goals and motivations.

Wyatt, O'Donnell, Beard and Busuttil (1999) in their analysis of 59 fatal Powered Two-Wheeler (PTW) crashes in Scotland drew the conclusion that post-crash medical attention was limited in effectiveness for preventing death. Rather, the greatest reduction in deaths of riders could be achieved by using crash reduction and injury prevention methods. One way of achieving this is by interventions that change rider behaviour, with an understanding of riders' attitudes to risk and enjoyment along with riding goals being an important part of any intervention design.

Intervention Targeting

Interventions are more likely to be successful and accepted by riders if the rider believes that the intervention is applicable to them and not just a generic solution that is more applicable to other road users (Broughton 2008a). Therefore interventions should address specific problems that are predominately an issue for particular sub-groups of riders and any marketing relating to such interventions ought to be designed to reach the specific groups being targeted.

The rider's view on how risk affects their riding enjoyment gives rise to three risk/enjoyment sub-groups: 'Risk Averse'; 'Risk Acceptors'; and 'Risk Seekers'(Broughton and Stradling 2005). 'Risk Seekers' are a small proportion of riders, but these riders may be over represented in the Killed or Seriously Injured (KSI) crash statistics because they get enjoyment from risk and may deliberately ride where task demand approaches, or exceeds, the limits of their capability; that is their enjoyment may be amplified by a level of anxiety.

Another item that is important within the design of an intervention is an understanding of why people ride. If feeling at risk is not an aim of riding for most riders, then appreciating the ways in which enjoyment is found in riding may offer more insight into the reasons for behaviour. This may allow the design of interventions that address the riskier elements of that behaviour without reducing enjoyment levels.

Broughton (2008a) found that riding enjoyment could be found in two ways: 'Rush Based Enjoyment' (RBE) and 'Challenge Based Enjoyment' (CBE). Generally, younger riders are biased towards RBE, while older riders tend to find more enjoyment from CBE.

These are generalities and individual enjoyment profiles for riders are complex, therefore profiling prior to any non-mass media interventions could be beneficial. The enjoyment profile differences between riders could also be considered for mass media intervention design, as one size does not necessarily fit all!

Respecting the Goals of Riding

The evidence, much of which has been presented in this book, does not support the idea that riders are the risk-taking outlaws that they were often portrayed as. Rather, for most riders the main goal of riding is enjoyment, which can be experienced as RBE or CBE. Enjoyment in riding is often found in an expressive element, even if there is a functional aspect to the ride, for example, some may use their bike to commute to work but part of the reason for doing so is the enjoyment gained from using this method of transport. The majority of riders know the risks involved in riding, yet in order to experience enjoyment from riding they are willing to accept this risk (Sexton et al. 2006) – that is, they ride despite the risk, not because of the risk (Broughton 2006).

A symptom of the pleasure riding goal is shown whereby crashes are happening with reference to the rider's domicile with riders significantly more likely to be crash involved at greater distance from home (see Table 2.8).

As enjoyment is a major riding goal, then any safety intervention must respect this. If an intervention fails to acknowledge, or attempts to remove, this goal, it is likely that riders will reject the intervention and it will be ineffective.

Updated Practical Test

The European Community (EU) has drawn up various Driving Licence Directives. These will have a big impact on how motorcycle licences are obtained. The UK Government is currently implementing the Second Driving Licence Directive (2DLD). The Direct Access Scheme (DAS) that allows certain riders an accelerated route to obtain an unrestricted licence, which is part of this Directive, was introduced in 1997. Further sections of the 2DLD are due to be introduced in October 2008 that will mean that part of the practical riding test will be taken off-road at Driving Standards Agency (DSA) 'super sites' (see Chapter 2 for more details of what is involved in this new practical riding test). One of the issues of concern is that the test centres where a motorcycle practical test can be taken will be significantly reduced across the UK, for example, the test centres within Scotland are due to reduced from 64 to 12 (Ross 2008). David Ross of *The Herald* newspaper reported that:

> ... for up to six months after the regulations come into force at the end of September, just a handful of centres will be open. To date, the Driving Standards Agency (DSA) has acquired four sites in Scotland: East Lothian and Glasgow, which are both ready; Kirkcaldy and Inverness, which are under construction.
>
> In addition, DSA currently expects Aberdeen will be operational when it all starts on September 29. It all represents a radical departure from the present provision which allows motorcycle tests in some of the most remote parts of the country from Arran to Ardnamurchan and Tiree to Thurso. (Ross 2008)

One of the fears being expressed in this article is that some riders who would face a very long trip to access a test centre will decide to ride illegally, and if a rider is willing to flout the law on holding a valid licence they may also be willing to flout the law on the power of the bike that they ride with respect to their age.

Skills Training

Training can aid in improving skills of both experienced and inexperienced riders; the resultant improvement in skills may seem to be the solution to the problem of rising KSI numbers. However, research shows that those who undergo further training are more likely to be at risk while using the roads (Rutter and Quine 1996). One effect that training may have is in reducing the number of convictions for traffic law offences that a motorcyclist may have; however, the number of crashes that trained riders have may not be different from untrained riders (Rothe and Cooper 1987). Rider skills training assumes that the rider's knowledge and skills are not of a sufficiently high standard or are faulty and thereby by increasing these skill levels will decrease the likelihood of a rider experiencing a crash (Shaoul 1975). Therefore the training will mean that the rider will perform better in traffic situations (Rothe and Cooper 1987). In the evaluation of the Bikesafe Scotland scheme, a significant number of those who took part rode harder out of town after they completed the course than they did prior to undertaking it. This may be due to riders perceiving their skills to have been enhanced and therefore feeling they have a higher capability (Ormston, Dudleston, Pearson and Stradling 2003). This may give rise to the situation so well described by Robert Gifford, executive director of the Parliamentary Advisory Council for Transport Safety (PACTS):

> Improving motorcyclists' skills merely means they kill themselves in a more skilled way. (Gifford 2003)

Training that riders can undertake on a voluntary basis may have little impact on reducing KSI rates. Rothe and Cooper (1987) found that those who decided to undertake additional driver training on a voluntary basis tended to be more deliberate and restrained and were less likely to be hurried in their actions compared to those who would not voluntary take up training.

If 'skills training' alone does not necessarily increase safety, then how can rider training be used to reduce the KSI rate and improve motorbike rider safety? One method is not just to focus upon skills. Hatakka, et al. identified a hierarchy of driver training with four levels (Hatakka, Keskinen, Gregersen, Glad and Hernetkoski 2002). The lowest level of the hierarchy is concerned with basic driving, or riding, skills, such as how to manoeuvre the vehicle. The second level deals with how the road user adapts to the changing demands of the road situation. The next two levels are more concerned with goals; level three being specifically about the goals of road use, such as the purpose of vehicle use. The goals of life and living are found in the fourth, and highest, level of the hierarchy where such items as the importance of the vehicle of choice are found. Therefore the lower two levels of this hierarchy are concerned with gaining mastery over the vehicle by learning how to manoeuvre it and how to adapt to the various demands placed upon the rider/driver by the ever changing road situation and the upper two levels of the hierarchy concern wider goals, the goals of driving and the goals of life.

Most pre-test training, such as the Compulsory Basic Training (CBT) that is required before a rider is allowed to use a motorised bike on the public road is focused on the lower two levels: bike control and reading/reacting to the traffic situation. Post-test training, such as 'Bike Safe', a scheme where riders are assessed by police motorcyclists (BikeSafe 2008; Ormston, Dudleston, Pearson and Stradling 2003) concentrates on the reading of other traffic. Accordingly it focuses mainly on the second lowest level of the hierarchy. Training schemes such as this, that focus on these levels can increase the vulnerability of riders by raising the perceived skill levels of riders (Goldenbeld, Twisk and de Craen 2004; Rutter and Quine 1996). This is not to say that training on these levels should not take place as the riding skills that are taught within these training environments are essential for safe riding, but training schemes need to temper the taught riding skills by also placing emphasis on the 'goals and context of riding'.

The effect that skills training can have on how a rider rides can be explained in terms of task difficulty as skills training has the aim of increasing a rider's capability. One of the effects of increasing the skill level of a rider is that this will also increase the threshold where task demand approaches capability due to an increase, or perceived increase, in capability. If a rider gains enjoyment from CBE then they may feel that they will have to ride harder to reach that threshold point. A basic level of skill is needed to ride, and it is not suggested that these, or more advanced skills, should not be taught. However, any skills-based training, needs to be designed to inoculate the rider against riding harder and faster because of the training. To accomplish this, psychological techniques could be employed to address the upper two levels of the hierarchy of driver training, which emphasises that the rider skill levels may be lower than they believe. Consideration could also be given to simple aspects of training, for example, the title of 'advanced training' may suggest to those who have undertaken this training that they are now highly skilled and their riding confidence may reflect this belief even if their riding competence does not.

Training could provide a rider with the aptitude to prevent a misjudgement of speed that could cause task demand to exceed capability (C<D). Assessment of the way that riders seek enjoyment from riding may assist in identifying appropriate training and how it should be implemented. Riders inclined to gain RBE, where speed is a main component, may benefit from training to give them the ability to correctly judge their ridden speed. As younger riders are more likely to gain enjoyment from RBE and they are over-represented in the KSI crash figures, speed and task demand awareness training for younger riders may be effective as a crash reduction strategy. Riders seeking CBE may benefit more from skills based on greater hazard perception and techniques designed to improve defensive riding – appealing to the challenge that can be found in developing safer riding techniques.

Some riders seek a flow type experience when riding by attempting to match their riding skills to the challenge presented by the riding environment. This is equivalent to task demand being closely matched to capability (Fuller 2005), which can be expressed as C≈D. When a rider is in this state then there is a very small safety margin and if task demand rises, or capability drops, (C<D) then a rider would be out of control which may result in a collision. Skills training may increase rider capability; therefore a higher task demand would be required to achieve a C≈D state. As speed is a major enhancer of task demand then it would be logical that skills training may entice riders who are seeking a flow experience to ride faster. Therefore, any skills-based intervention should look at inoculating against this phenomenon (Mannering and Grodsky 1995; Ormston, Dudleston, Pearson and Stradling 2003).

Skills training is essential in attempting to reduce KSI crashes by improving skills but as this may result in making the rider feel they are more capable, it may only result in 'harder' riding and have little, or even a negative, impact on reducing crashes. Therefore training should seek to change behaviour by addressing the two high levels of the Driving Training Hierarchy described by Hatakka et al. (2002) of rider goals and goal of life.

Behavioural Change

In order for training to be successful, behaviour and how it may be influenced has to be understood. Lewin (1935) suggested that behaviour is driven by the characteristics of the person and the environment. Therefore behaviour (B) can be expressed as a function of the interaction between the environment (E) and the individual characteristics of a person (P), or $B = f(P, E)$ (Lewin 1935). As behaviour is an interaction of individual characteristics with the physical and social situation, then it can be seen to be both complex and dynamic in nature. This does suggest, however, that we can change the behaviour of a particular person with a particular environment by adjusting their individual characteristics. This, therefore, gives a potential method of making riders ride in a safer manner.

One such intervention that is being developed is a rider improvement course that concentrates on attitudes and beliefs rather than skills (Burgess 2005). A course, similar to the Driver Improvement Scheme (Broughton 2008b; Burgess 2005; Stradling and Cunliffe 2006) is being piloted in 2008/9. This course, designated RIDE (Rider Improvement Delivering Experience), is a diversion from prosecution course where riders take the course in place of paying a fixed penalty fine and having penalty points placed on their licence. This gives the benefit of getting those with a high likelihood of becoming a KSI statistic onto the course, and often these are the riders that are less likely to take up any sort of voluntary motorcycle training. The RIDE course is based around the Transtheoretical Model (Prochaska, DiClemente and Norcross 1992), which is more commonly known as the Stages of Change Model. The Stages of Change Model proposes that people go through various stages on the way to behaviour change and therefore interventions can be designed to help them to move between these stages. This model has been used to modify various behaviours, such as helping smokers to quit. The idea is to get people to move from a position where they resist change, through to thinking about it, planning to make changes, making the changes and then implementing methods to try and make the changes permanent and prevent relapse. The RIDE course implements a version of this model by first convincing riders that there is a real problem with motorcycle road safety, then riders are shown that they are part of the problem. Using examples, riding myths are dispelled such as the level of protection that motorcycling equipment provides and how good, or bad, at hazard perception the human is. Riders are also invited to think about what the consequences are to their friends and family if they do not return from a ride. The final part of the course focuses on the final part of the Transtheoretical Model, maintenance; that is, how riders can follow through on their interventions after the course is over. The course is designed to address issues of maintenance through devices such as getting the rider to make a pledge of the changes that they intend to make (Owl Research Ltd 2008). This type of formalised 'commitment' can help focus the rider on longer-term changes. Earlier, in Chapter 9, the impact of groups and peer pressure on riders was discussed. While much of this focussed on the negative impacts that may occur when riders seek to conform to the perceived 'norms of the group', this peer pressure and conformity to norms of behaviour can also be used as a positive force.

Using Group Pressure for Behavioural Change

One of the tools that the RIDE courses uses is group discussion between the riders on the course. Within the course some attendees will have beliefs that are contrary to safe riding and having the evidence presented to show that these are not correct may not be sufficient to overturn these beliefs. Within the group discussion environment, a minority who are not initially willing to change their incorrect beliefs may be willing to do so in order to go along with the majority (Asch 1956). This, in part, will be because the group has not validated their beliefs, so they will adopt the group belief as being the correct one (Festinger 1953). They may also fear the ridicule of other riders for holding beliefs that, although thought to be part of the social norm for riders, are shown not be the social norms of the group of riding peers, all of which are on the course because they have contravened traffic law (Deutsch and Gerard 1955). Therefore conformity can be driven by a group goal (Festinger 1950) and, if this goal is manoeuvred to be one of achieving safer riding practices, then there is a higher probability of successfully changing rider behaviour for the better (Brown 2000). If the whole group, via a discussion, agree on safer riding goals, then there is a better chance of success than if someone just lectures them (Lewin 1965). By getting the majority of the group to adopt beliefs that may lead to safer riding, then group pressure can be used to restrain anti-social, or unsafe, riding behaviour (Milgram 1964).

The group discussion is also a useful tool if one of the riders on the course shows a particularly deviant attitude, as the others are liable to isolate that rider thus demonstrating that his social norms are not in line with the group's social norms (Allen and Wilder 1980).

The use of interventions based on the Transtheoretical Model or around positive peer pressure are not the only ways to get inside the head of a rider and adjust their riding behaviour. As the kinaesthetic nature of riding resonates with many sporting activities the methods from sports coaching can also be used as the basis for ways to improve riding.

Sports Coaching Techniques

Motorcycle riding has similarities with sports participation as it is often done as an expressive activity, it requires cognitive and motor expertise to master the activity and there may also be elements of competition between riders. Therefore some of the techniques used within sports coaching may be valid in motorcycle safety intervention design

Motivation

What motivates a sportsman? Achievement is often a major factor. Atkinson (1964) proposed that the motivation for achievement was a combination of two components: the 'motive to approach success' and the 'motive to avoid failure'. He suggested that everyone has both motivations, but not to the same degree. The motivation of success or failure is evaluated by the goals that the individual has set for themselves (Gill 1986).

Cogan and Brown (1999) reported that those who take part in 'risk sports' may not decide to pursue these activities because of the risk or the emotions that are invoked by the risk but rather the involvement in the risky activity is related to mastery and gaining control over their environment (Hatzigeorgiadis 2002). This is the case for the majority of riders with the seeking of risk not being the aim of riding, but rather the gaining of mastery of the riding task. This is related to using the skills that a rider has to match the challenge of the riding environment, with speed being the main method of adjusting the level of challenge (Fuller 2008; Hannigan, Fuller, Bates, Gormley, Stradling, Broughton, Kinnear and O'Dolan 2008).

The understanding of the motivation to ride shows that it may not be profitable to try to alter the way riders view the risk of riding (Sexton, Hamilton, Baughan, Stradling and Broughton 2006), but rather how they seek challenge (Broughton 2008a).

Perceived control is also an important factor for sports motivation, and again this is related to riders needing to be in control (Broughton and Stradling 2005). If a sports person is given even a small increase in the amount of choice that they have, this often translates into positive motivational and behavioural changes (Thompson and Wankel 1980). Thus if we wish to increase the chances of a rider being motivated to strive to change their behaviour, they must perceive that they have some choice in how they accomplish this – preaching a 'fixed method' is less likely to result in effective change in rider behaviour.

The use of sports psychology methods is not just limited to understanding the whys of riding; it can also be used to develop skills such as hazard perception by giving insight into how a rider may perceive surroundings when riding.

Attention

Hazard perception is of vital importance for riders on the public road (Wallace, Haworth and Regan 2005) and therefore attention is a significant skill. There is a capacity limit to attention, and within sport this has to be taken into account. These limits to 'control processing' can be overcome by moving the skill to automatic processing, which is not limited by attentional capacity (Gill 1986). Practice is important for skills to become automatic, with those who practise more at a specific sport being better at recalling game situations (Allard, Graham and Paarsalu 1980). Practice can also allow players to pick up advanced cues that allow them to predict what is about to happen and react to it (Andrew 1989; Tenenbaum and Lidor 2005). Therefore rehearsing actions assists in improving performance; rehearsal though is only useful if it is being done correctly. It is the job of the coach to ensure that this is the case by providing fast and accurate feedback to the sportsman (Gill 1986).

Riders can be taught how to carry out hazard perception, and with practice these skills can be habitual, even to the level where schema can be triggered that alerts the rider of potential threats. However, the practice for hazard perception has to be carried out correctly otherwise the rider may be habitualising incorrect skills such as ineffectual visual search patterns. Only very rudimental hazard perception skills can be automated such that they run a schema of automated response, for example, reacting to seeing a brake light in front illuminate. The majority of hazard perception has be done on a conscious cognitive level (Horswill and McKenna 2004), but a well honed set of schema can alert the conscious of an impending threat (Broughton 2006).

Coaching

Sports personnel are coached so that their skill levels, and performance, improve. Motorcycle safety interventions have a similar goal, to improve the safe-riding skills and performance of riders.

There are two basic sports coaching behaviours: reactive, responding to the sportsman's behaviour and actions; and proactive, where the coach instigates the coaching (Smith and Smoll 1977), but regardless of which style is used the important aspect is to ensure that bad habits are corrected before they become automated.

The fours Cs are often considered as the main pillars of the mental qualities that are required for an athlete to be successful (Mackenzie 2007):

1. concentration
2. confidence
3. control (keeping emotional control)
4. commitment.

The skill of imagery can aid in the above. Imagery is when the sportsman imagines that they are performing the skill correctly and well. This technique, when practised correctly, can aid in increasing performance and the learning of skills (Gill 1986). Other methods can be used, such as self-talk. Self-talk is a method that can be used while the athlete is participating in sport. It is the sportsman talking to himself, repeating a mantra about his performance, such as 'keep the feet moving'. Self-talk can also be negative and this can undermine confidence and act as a distraction, therefore lowering the level of control (HarrowDrive 2006) but positive self-talk is a means to keep attention focused and to help to overcome bad habits (Williams and Leffingwell 1996).

Performance of athletes can be improved by coaching that teaches correct techniques and provides opportunities to use those skills to the best capability of the sportsman.

Both of the above methods can be used to improve riding techniques. A rider may be asked to imagine correctly, and safely, riding around a corner – choosing the correct line and speed. By repeatedly doing this the mental system will remember the cues and in a real world situation run these in a similar situation. Self-talk can also be used with a rider repeating a mantra about a certain part of the riding task, such as 'slow into a corner, fast out'. However, this method must be used with care as the action of self-talk adds an extra task to the riders and thus increases task difficulty. Both of these methods are used to aid in placing skills into procedural memory.

Procedural memory is sometimes referred to as 'muscle memory', even though the muscle has no memory, (Gill 1986) as it often seems that the muscles know what actions to take without any input from the conscious, or explicit, memory. Within sport it is often important that athletes quickly and correctly respond to stimuli by executing motor actions. This primed reaction relies on learning to associate, with practice, the stimuli with the execution of the correct motor movements. This association is stored in the implicit memory allowing the learnt actions to be carried out in an automatic, fast and efficient manner (Kibele 2006; Zeigler 2002).

The idea of practising skills so that they become automatic is one that is often found in sport (Hogarth 2001; Raab 2003). For example, the skill of batting in cricket is very complex, with a batsman having to decide what trajectory the ball is moving along, where and how much it will bounce, whether to move forward to the ball or back, where the fielders are positioned, what shot to play and then to execute the motor actions required to play this shot (Andrew 1989). A batsman often has less than a second from when the ball leaves the bowler's hand until the ball has passed him. For a batsman to carry out the above processing using explicit memory would be too slow, hence it has to be carried out automatically using implicit or procedural memory (Broughton 2006; Kibele 2006). For a complex skill such as batting to be learnt it has to be broken down into smaller sub-sets of skills that can be initially carried out explicitly. With practice these skills will eventually be moved from explicit memory to the implicit and carried out automatically. Once this has been accomplished then the next skill sub-set can be taught.

Riding, like batting at cricket, is difficult and complex, involving a range of movements and responses. Therefore it can be expected to be treated like an implicit skill. This is why lessons from sports coaching, and how sportsmen train, can be applied to riding.

Teaching such techniques during training may assist in longer-term re-enforcement of skills taught. As with sportspeople, most riders need some encouragement to maintain good habits, therefore regular assessment and re-alignment of behaviours is necessary. While the fiscal implications of further testing may be an issue, given the relatively high KSI crashes experienced by this road group, it may be justified. There may be possibilities of using computer technology to re-assess trained riders after a period of time. Programmes similar to those used in psychology to modify thinking patterns and behaviour, such as cognitive behavioural therapy, could be utilised to help riders ride in a safe manner.

With new skills should come a way for riders to assess that the new skills are being used correctly (self assessment). With helmet cameras becoming cheaper and more accessible it is now practical for riders to record their rides and then for the footage to be reviewed by the rider to evaluate it against what they had been taught. However, equipping riders with helmet cameras may also cause problems due to audience effect by proxy, resulting in showing off by riding in a way that they feel will be acceptable to their peers.

As discussed above, techniques used in sports coaching can be applied to improving and developing safe riding habits. The visualisation technique allows for a sportsman to rehearse in absentia being in a specific position so that when they find themselves in that situation for real they react correctly. Within sport, rehearsal helps players pick up advanced cues to what is happening (Andrew 1989), and this remains true for riding. Learning hazard perception and reacting correctly to these hazards is important for riding and a proficiency that should be taught to all riders. It is especially important as, if a rider has to react unexpectedly to a hazard, this can reduce capability at the same time as task demand increases, a double whammy that increases the chance of being out of control.

These types of interventions are focused on the rider and seek to adjust the characteristics of the individual. However, Lewin (1935) indicated that behaviour is a function of both personal characteristics and the environment. Therefore to change behaviour there may be opportunities to adapt the environment experienced by the rider.

Non-rider Based Interventions

The rider environment consists of the static physical aspects such as road surfaces, more fluid aspects such as road conditions and the attitudes and subsequent behaviour of other road users. Interventions to improve safety could be developed from any of these perspectives.

Environmental interventions can take a number of guises, for example, road engineering interventions could impact on the static physical environment by improving street furniture from the perspective of the rider. An example of this is an engineering intervention implemented by Buckinghamshire County Council where they extended the road markers around a bend, the idea being that if the rider's eye is held around a bend then there is a higher chance that it will be successfully negotiated (Institute of Highway Incorporated Engineers 2005).

Although some road conditions are a result of unavoidable situational factors such as poor weather reducing grip on the road, other elements affecting road conditions such as diesel spills can be at least reduced through enforcement of legislation that allows prosecution of those who spill diesel as well as more efficient clean up of spills, but this would require heightened priority of such operations.

Interventions that change rider behaviour indirectly, such as allowing riding in bus lanes, can also be assessed using the task difficulty model by determining if the engineering or environmental change will affect the task difficulty, and how riders will react to this change. The design of these types of interventions needs to take into account the reaction of riders, including changing goals and reactions to risk. For example, in areas where a rider can see an opportunity for enjoyment, there may be a temptation to ride hard, but some road features can make a road look risky (see Chapter 7). Therefore if a road is engineered so as to look risky and not tempting, then a rider is likely to ride more carefully.

The hazards posed by other road users being unaware of PTW users could be improved through awareness raising, such as the 'think' campaigns discussed in Chapter 2 (DfT 2006e) or through additional driver training.

SMIDSY – Educating Non-riders

A significant number of crashes involving motorcycles are caused by other road users, with these often having violated the PTW's right of way (Clarke, Ward, Bartle and Truman 2004). For example, a car may pull out of a junction or make a U-turn without fully checking and pull out into the path of a PTW. Often in these situations the driver claims that they did not see the oncoming motorcycle – a looked but failed to see (LBFTS) error. This is frequently referred to by riders as a SMIDSY standing for 'Sorry Mate I Didn't See You' as these are often the first words spoken by the driver to the fallen rider (South Gloucestershire Council 2008).

It has been stated that those who ride and drive, or have close friends or family that ride, are more likely to perceive the oncoming motorcycle (Brooks and Guppy 1990; Magazzu, Comelli and Marinoni 2006). This suggests that the best way to get drivers to detect oncoming motorcycles is to get them to ride one; however, this is impracticable and also probably politically unsavoury. Therefore means to get drivers to gain a similar experience is required so that drivers understand the hazards that bikes experience (Crundall, Bibby, Clarke, Ward and Bartle 2008).

Summary

As PTW riders are a vulnerable road user group, interventions for their safety are needed. This book has shown that rider goals and motivations are different from car drivers so generic road safety solutions may not be effective. For any intervention to be effective it must be designed specifically around rider goals and not the goals that the intervention designers and policy makers believe riders have. Interventions need to be built around the principle that an appreciable number of riders use their PTWs for expressive riding with enjoyment, not risk seeking, being the main goal. Therefore interventions must be built around safe enjoyable riding rather than trying to convert riders from the stereotype of 'risk junkies', which is rarely a true depiction. As the majority of riders do not ride because of the risk, most would accept interventions that reduce the risk provided there was not a significant erosion of enjoyment.

Riders ride because they enjoy it, finding enjoyment in a combination of rush and challenge. Both of these elements need to be considered when interventions for PTWs are being designed. Further to this, riders also view risk in three distinct ways, some not enjoying risk, some accepting a level of risk to gain enjoyment and a small minority who enjoy risk. It is theorised that riders who seek risk are more likely to be involved in a crash, and therefore could be specifically targeted by safety interventions. The main aspect of any intervention must respect the goal of enjoyment – attempting to remove this goal will only alienate those the intervention is designed to help.

References

Aarts, L. and Van Shagen, I. 2006. Driving speed and the risk of road crashes: A review. *Accident Analysis and Prevention*, 38, 215–24.

ACEM. 2000. *Smart Wheels for City Streets*. Brussels: ACEM.

ACEM. 2004. *Maids: In-depth Investigations of Accidents Involving Powered Two Wheelers*. Brussels: ACEM.

Adams, J. 1995. *Risk*. New York: Routledge.

Allard, F. 2001. *Kinesiology 356 Course Notes*. Waterloo, Ontario: University of Waterloo.

Allard, F., Graham, S., and Paarsalu, M.T. 1980. Perception in sport: Basketball. *Journal of Sport Psychology*, 2, 14–21.

Allen, V.L. and Wilder. D.A. 1980. Impact of group consensus and social support on stimulus meaning: mediation of conformity by cognitive restructuring. *Journal of Personality and Social Psychology*, 39, 1116–24.

Alt, J. 1982. Popular culture and mass consumption: The motorcycle as cultural commodity. *Journal of Popular Culture*, xv(4), 129–41.

American College of Surgeons. 2004. Statement of support of motorcycle helmet laws. *Journal of the American College of Surgeons*, 199(2), 260.

Andrew, K. 1989. *The Skills of Cricket*. Marlborough, Wiltshire: Crowood Press Ltd.

Argyle, M. and Hills, P. 2000. Religious experiences and their relationships with happiness and personality. *International Journal for the Pschology of Religion*, 10, 157–72.

Arnett, J. 1994. Sensation seeking: A new conceptualization and a new scale. *Personality and Individual Differences*, 16(2), 289–96.

Asch, S.E. 1956. Studies of independence and conformity: A minority of one against a unanimous majority. *Psychological Monographs*, 70(a), 1–70.

Atkinson, J.W. 1964. *An Introduction to Motivation*. Princeton, NJ: Van Nostrand.

Ball, S. and Zuckerman, M. 1992. Sensation seeking and selective attention: Focused and divided attention on a dichotic listening task. *Journal of Personality and Social Psychology*, 5, 825–31.

BBC. 2004. *Fatal Motorcycle Crashes Increase*. [Online]. Available at: http://news.bbc.co.uk/1/hi/england/northamptonshire/3097267.stm [accessed: 12 February 2005].

BBC. 2005. *The Archers: Alan Franks*. [Online]. Available at: http://www.bbc.co.uk/radio4/archers/whos_who/characters/alan_franks_no_js.shtml [accessed: 28 October 2007].

BBC. 2006. *Bikers Meet Over 'Park Ban' Call.* [Online]. Available at: http://news.bbc.co.uk/1/hi/wales/north_east/4790131.stm [accessed: 17 September 2007].

BBC. 2007a. *1964: Mods and Rockers Jailed After Seaside Riots.* [Online]. Available at: http://news.bbc.co.uk/onthisday/hi/dates/stories/may/18/newsid_2511000/2511245.stm [accessed: 4 June 2007].

BBC. 2007b. *Armed Biker in Raid on Jewellers.* [Online]. Available at: http://news.bbc.co.uk/1/hi/england/essex/7052214.stm [accessed: 4 March 2008].

BBC. 2007c. *Armed Biker Takes Church Donation.* [Online]. Available at: http://news.bbc.co.uk/1/hi/england/kent/7010342.stm [accessed: 4 March 2008].

BBC. 2007d. *Average Earnings at £457 per Week.* [Online]. Available at: http://news.bbc.co.uk/1/hi/business/7082630.stm. [accessed: 5 May 2008].

BBC. 2007e. *Bikers' Easter Joy for Sick Kids.* [Online]. Available at: http://news.bbc.co.uk/1/hi/scotland/glasgow_and_west/6530139.stm [accessed: 4 March 2008].

BBC. 2007f. *Brotherhood with History of Bad Blood.* [Online]. Available at: http://news.bbc.co.uk/1/hi/uk/6944876.stm [accessed: 5 March 2008].

BBC. 2007g. *Close Shave Earns Man Driving Ban.* [Online]. Available at http://news.bbc.co.uk/1/hi/scotland/tayside_and_central/6335491.stm [accessed: 6 February 2007].

BBC. 2008a. *Biker Charity Offers NHS Delivery.* [Online]. Available at: http://news.bbc.co.uk/1/hi/england/merseyside/7202427.stm [accessed: 15 June 2008].

BBC. 2008b. *Biker Club 'in Channel Islands'.* [Online]. Available at: http://news.bbc.co.uk/1/hi/world/europe/jersey/7052215.stm [accessed: 21 July 2008].

BBC. 2008c. *Biker's 189mph Stunt is Condemned.* [Online]. Available at: http://news.bbc.co.uk/1/hi/england/gloucestershire/7188000.stm [accessed: 1 August 2008].

BBC. 2008d. *Four in 10 Motorbikes 'Not Taxed'.* [Online]. Available at: http://news.bbc.co.uk/1/hi/uk_politics/7200066.stm [accessed: 22 June 2008].

BBC. 2008e. *The Hairy Bikers' Cookbook.* [Online]. Available at: http://www.bbc.co.uk/food/tv_and_radio/hairybikers_about.shtml [accessed: 15 May 2008].

Beck, C., Wets, G., Torfs, R., Mensink, C., Broek, S. and Janssens, D. 2006. Impact of trip purpose on driving behaviour: Case study on commuter traffic in Belgium, in *International Symposium Transport and Air Pollution: Vol. 15.* Reims: France.

Bellaby, P. and Lawrenson, D. 2001. Approaches to the risks of riding motorcycles: Reflections on the problem of reconciling statistical risk assessment and motorcyclists' own reasons for riding. *Sociological Review*, 49(3), 368–88.

BikeSafe. 2008. *Welcome to the Official BikeSafe website.* [Online]. Available at: http://www.bikesafe.co.uk/ [accessed: 15 June 2008].

Blanchard, H.T. and Tabloski, P.A. 2006. Motorcycle safety: Educating riders at the teachable moment. *Journal of Emergency Nursing*, 32(4), 330–2.

BMF. 2004. *BMF Diesel Clean-up Campaign.* [Online]. Available at: http://www.bmf.co.uk/diesel_spills/index.html [accessed: 19 February 2007].

BMF. 2005. *Diesel Spills and Motorcyclists.* [Online]. Available at: http://www.bmf.co.uk/briefing/dieselspillsandmotorcyclists.html [accessed: 19 February 2007].

BMF. 2007 *Manifesto.* [Online]. Available at: http://www.bmf.co.uk/pages/bmf_main_pages.php?main_page_id=152 [accessed: 2 March 2008].

BMF. 2008. *About BMF.* [Online]. Available at: http://www.bmf.co.uk/pages/bmf_main_pages.php?area_id=5 [accessed: 2 March 2008].

Bradley, H. 1996. *Fractured Identities: Changing Patterns of Identity.* Cambridge: Polity.

Branas, C. and Knudson, M. 2001. Helmet laws and motorcycle rider death rates. *Accident Analysis and Prevention*, 33(5), 641–8.

Brickman, P. and Campbell, D.T. 1971. Hedonic relativism and planning the good society, in *Adaptation Level Theory: A Symposium*, edited by M.H. Appley. New York: Academic Press.

British Medical Journal. 2002. *The Safest Cars?* [Online]. Available at: http://www.bmj.com/cgi/eletters/324/7346/1149 [accessed: 15 June 2006].

British Standards. 2002. *Protective Clothing for Professional Motorcycle Riders. Jackets, Trousers and One Piece or Divided Suits. General Requirements.* [Online]. Available at : http://www.standardsdirect.org/standards/standards1/StandardsCatalogue24_view_10883.html [accessed: 19 June 2007].

British Standards. 2003. *Motorcyclists' Protective Clothing Against Mechanical Impact. Motorcyclists Back Protectors. Requirements and Test Methods.* [Online]. Available at: http://www.standardsdirect.org/standards/standards1/StandardsCatalogue24_view_12112.html [accessed: 19 June 2007].

Broadbent, D.E. 1958. *Perception and Communication.* New York: Pergamon Press.

Brooks, P. and Guppy, A. 1990. Driver awareness and motorcycle accidents. *Proceedings of the International Motorcycle Safety Conference,* 2[10], 27–56.

Broughton, P.S. 2005. Designing PTW training to match rider goals, in *Driver Behaviour and Training, Vol 2*, edited by L. Dorn. Aldershot: Ashgate Publishing.

Broughton, P.S. 2006. The implication of the flow state for PTW training in *Behavioural Research in Road Safety 2006, Sixteenth Seminar.* London: DfT.

Broughton, P.S. 2007. *Risk and Enjoyment in Powered Two Wheeler Use.* Unpublished PhD thesis, Transport Research Institute, Napier University.

Broughton, P.S. 2008. *RIDE.* [online]. Available at: http://www.ride-course.org/about.html [accessed: 30 August 2008].

Broughton, P.S. 2008a. Flow, task capability and Powered-Two-Wheeler (PTW) rider training in *Driver Behaviour and Training, Vol 3*, edited by L. Dorn. Aldershot: Ashgate Publishing.

Broughton, P.S. and Stradling, S. 2005. Why ride powered two wheelers? in *Behavioural Research in Road Safety 2005, Fifteenth Seminar*. London: DfT.

Brown, R. 2000. *Group Processes: Dynamics Within and Between Groups.* Oxford: Blackwell Publishers Ltd.

BSB. 2006. *Virgin Mobile Yamaha*. [Online]. Available at: http://www.britishsuperbike.com/teams.cfm?team_id=5 [accessed: 4 February 2007].

Burgess, C.N.W. 2005. *Motorcycle Offender Re-Education: Injury Reduction or Recidivism Prevention?* Paper to the: 2nd International Conference on Driver Behaviour and Training, Edinburgh, 15 November 2005.

Burton, H. 1954. Most unpopular men on the road. *Saturday Evening Post*. 25 September 1954, 33.

Business Week. 1965. The double life pays off. *Business Week*. 1 July, 24, 22.

Central Office of Information for Department of Transport. 1978. *Think Bike – Jimmy Hill*. [online]. Available at: http://www.nationalarchives.gov.uk/films/1964to1979/filmpage_jimmy.htm [accessed: 12 August 2007].

Chesham, D.J., Rutter, D.R. and Quine, L. 1993. Motorcycling safety research: A review of the social and behavioural literature. *Social Science Medicine*, 37(3), 419–29.

Chinn, B. and Hopes, P. 1985. Leg protection for riders of motorcycles in *Proceedings of the 10th ESV Conference*. Oxford: ESV.

Chorlton, K. and Jamson, S.L. 2003. Who rides on our roads? An exploratory study of the UK motorcycling fleet, in *Behavioural Research in Road Safety: Thirteenth Seminar Proceedings*. London: DfT.

Church, R.M. 1968. Applications of behaviour theory to social pyschology: Imitation and competition in *Social Facilitation and Imitative Behaviour*, edited by E.C. Simmer, R.A Hope and G.A. Milton. Boston: Allyn and Bacon.

City of York Council. 2005. *City of York Local Transport Plan 2006–2001. Annex H – Powered Two Wheelers Strategy*. York: City of York Council.

Clark, D.D. and Ward, P. 2002. Sequential case-study of police road accident files involving young drivers, motorcycles, or work related accidents, in *Behavioural Research in Road Safety: Twelfth Seminar*. London: DfT.

Clarke, D.D., Ward, P., Bartle, C. and Truman, W. 2004. *In-depth Study of Motorcycle Accidents*. London: Department for Transport.

Clarke, D.D., Ward, P., Truman, W. and Bartle, C. 2003. Motorcycle accidents: Preliminary results of an in-depth case-study. *Accidents in Behavioural Research in Road Safety: Thirteenth Seminar*. London: DfT.

Clutch and Chrome. 2008. *Hells Angels Motorcycle Gang Celebrations Pass Without Trouble*. [Online]. Available at: http://www.clutchandchrome.com/News/0803/News0803121.htm [accessed: 15 August 2008].

Cogan, N. and Brown, R. 1999. Metamotivational dominance states and injuries in risk and safe sports. *Personality and Individual Differences*, 27, 503–18.

Colbourn, C.J. 1978. Perceived risk as a determinant of driver behaviour. *Accident Analysis and Prevention*, 10(2), 131–41.

Committee of Public Accounts. 2008. *Evasion of Vehicle Excise Duty*. [Online]. Available at: http://www.parliament.uk/parliamentary_committees/committee_of_public_accounts/pacpn080122.cfm [accessed: 21 January 2008].

Coomber, R. 2004. *Using the Internet for Survey Research.* [Online]. Available at: http://www.socresonline.org.uk/socresonline/2/2/2.html [accessed: 5 December 2005].

Coombs, C.H., Donnell, M.L. and Kirk, D.B 1978. An experimental study of risk preferences in lotteries. *Journal of Experimental Psychology: Human Perception and Performance*, 4(3), 497–512.

Cornwall County Council. 2004. *Draft Cornwall Strategy for Powered Two Wheelers.* Truro, Cornwall: Cornwall County Council.

Cossack Owners Club. 2008. *Bikers and Image.* UK: Cossack Owners Club.

Cottrell, N.B. 1968. Performance in the presence of other human beings: Mere presence, audience and affilation effects, in *Social Facilitation and Imitative Behaviour*, edited by E.C. Simmer, R.A. Hope and G.A. Milton. Boston: Allyn and Bacon.

Crundall, D., Bibby, P., Clarke, D.D., Ward, P. and Bartle, C. 2008. Car drivers attitude towards motorcyclists: A survey. *Accident Analysis and Prevention*, 40, 983–93.

Csikszentmihalyi, M. 1997. *Finding Flow: The Psychology of Engagement With Everyday Life.* New York: Basic Books.

Csikszentmihalyi, M. 1990. *Flow : The Psychology of Optimal Experience.* New York: Harper & Row.

Csikszentmihalyi, M. 2000. *Beyond Boredom and Anxiety: Experiencing Flow in Work and Play.* San Francisco: Jossey-Bass.

Csikszentmihalyi, M. and Csikszentmihalyi, I.S. 1988. *Optimal Experience: Psychological Studies of Flow in Consciousness.* Cambridge: Cambridge University Press.

Csikszentmihalyi, M. and Hunter. J. 2003. Happiness in everyday life: The uses of experience sampling. *Journal of Happiness Studies*, 4, 185–99.

Cunneen, C. 1999. Zero tolerance policing and the experience of New York City. *Current Issues in Criminal Justice*, 10(3), 299–313.

Damasio, A.R. 2003. *Looking for Spinoza: Joy, Sorrow and the Feeling Brain.* London: Heinemann.

Deci, E.L. and Ryan, R.M. 1985. *Intrinsic Motivation and Self-determination in Human Behavior.* New York: Plenum.

Deutsch, M. and Gerard. H.B. 1955. A study of normative and informational social influence upon individual judgement. *Journal of Abnormal and Social Psychology*, 51, 629–36.

DfT. 2002. *Motorcyclists' Visors: Minimum Light Transmittance – Consultation Paper.* [Online]. Available at: http://www.dft.gov.uk/consultations/archive/2002/mv/motorcyclistsvisorsminimumli1736?page=1 [accessed: 3 July 2007].

DfT. 2003a. *National Travel Survey.* London: DfT.

DfT. 2003b. *Scoping Study on Motorcycle Training.* London: DfT.

DfT. 2004a. *Compendium of Motorcycling Statistics.* London: DfT.

DfT. 2004b. *The Highway Code.* London: DfT.

DfT. 2004c. *Tomorrow's Roads: Safer for Everyone – The First Three Year Review.* London: DfT.

DfT. 2005a. *The Government's Motorcycling Strategy.* London: DfT.

DfT. 2005b. *THINK! Key Facts and Research.* London: DfT.

DfT. 2006a. *Compendium of Motorcycling Statistics 2006.* London: DfT.

DfT. 2006b. *Driving Eyesight Requirements.* [Online]. Available at: http://www.direct.gov.uk/en/Motoring/LearnerAndNewDrivers/LearningToDriveOrRide/DG_4022529 [accessed: 18 November 2007].

DfT. 2006c. *National Travel Survey 2006.* London: DfT.

DfT. 2006d. *Theory Test.* [Online]. Available at: http://www.direct.gov.uk/en/Motoring/LearnerAndNewDrivers/TheoryTest/index.htm [accessed: 15 January 2007].

DfT. 2006e. *THINK! – Motorcycle Safety.* [Online]. Available at: http://www.thinkroadsafety.gov.uk/campaigns/motorcycles/motorcyclesmedia.htm [accessed: 15 January 2007].

DfT. 2007a. *Road Statistics: Traffic, Speeds and Congestion.* London: DfT.

DfT. 2007b. *About Compulsory Basic Training (CBT).* [Online]. Available at: http://www.direct.gov.uk/en/Motoring/LearnerAndNewDrivers/RidingMotorcyclesAndMopeds/DG_4022430 [accessed: 28 February 2008].

DfT. 2007c. *Compendium of Motorcycle Statistics 2007.* London: DfT.

DfT. 2007d. *Motorcycle Licence Requirements.* [Online]. Available at: http://www.direct.gov.uk/en/TravelAndTransport/Highwaycode/DG_069867 [accessed: 28 February 2008].

DfT. 2007e. *SHARP, The Helmet Safety Scheme.* [Online]. Available at: http://sharp.direct.gov.uk/ [accessed: 17 February 2008].

DfT. 2007f. *Transport Trends 2006.* London: DfT.

DfT. 2008. *Vehicle Excise Duty Evasion 2007.* London: DfT.

Diehm, R. and Christine, A. 2004. Surfing: An avenue for socially acceptable risk-taking, satisfying needs for sensation seeking and experience seeking. *Personality and Individual Differences*, 36, 663–77.

Diekmann, A. 1996. *External Costs and Benefits of Powered Two-wheelers.* Brussels: Association des Constructeurs Européens de Motorcycles (ACEM).

Dietrich, A. 2004. Neurocognitive mechanisms underlying the experience of flow. *Consciousness and Cognition*, 13(4), 746–61.

DirectGov. 2008. *Clothing and Weather Protection.* [Online]. Available at: http://www.direct.gov.uk/en/Motoring/LearnerAndNewDrivers/RidingMotorcyclesAndMopeds/DG_4022434 [accessed: 1 September 2008].

Dirks, T. 2006a. *Easy Rider (1969).* [Online]. Available at: http://www.filmsite.org/easy.html [accessed: 25 May 2006].

Dirks, T. 2006b. *The Wild One (1953).* [Online]. Available at: http://www.filmsite.org/wild.html [accessed: 25 May 2006].

DSA. 2006. *Motorcycle Practical Test Explained.* [Online]. Available at: http://www.dsa.gov.uk/Category.asp?cat=250 [accessed: 30 June 2007].

DSA. 2006. *Vehicle Requirements for Motorcycles and Mopeds.* [Online]. Available at: http://www.dsa.gov.uk/Bikes.asp?cat=118 [accessed: 29 June 2007].
Dulaney, W.L. 2005. A brief history of 'outlaw' motorcycle clubs. *International Journal of Motorcycle Studies*, (1), 3.
DVLA. 2006. *Riding a Motorcycle – The Licensing Position From 1 February 2001.* [Online] Available at: http://www.dvla.gov.uk/drivers/rdmcycle.htm [accessed: 30 June 2007].
Easterbrook, J.A. 1959. The effect of emotion on cue utilization and the organization of behavior, 66, 183–201.
Easterlin, R.A. 1995. Will raising the incomes of all increase the happiness of all? *Journal of Economic Behavior and Organization*, 27, 1–34.
Easterlin, R.A. 2005. Feeding the illusion of growth and happiness: A reply to Hagerty and Veenhoven. *Social Indicators Research*, 74(3), 429–43.
Elliott, M.A., Baughan, C.J., Broughton, J., Chinn, B., Grayson, G.B., Knowles, J., Smith, L.R., Simpson, H. 2003. *Motorcycle Safety: A Scoping Study. TRL Report 581*. Crowthorne: Transport Research Laboratory.
Elvik, R., Christensen, P. and Olsen, S. 2003. *Daytime Running Lights: A Systematic Review of Effects on Road Safety.* Oslo, Norway: Institute of Transport Economics.
Elvik, R. and Vaa, T. 2004. *Handbook of Road Safety Measures*. Amsterdam: Elsevier.
ETSC. 1998. *Forgiving Roadsides.* Brussels: European Transport Safety Council.
EuroRap. 2004. *EuroRAP 2004 British Results*. London: The AA Motoring Trust.
Evans, L. 1986. Risk homoeostasis theory and traffic accident data. *Risk Analysis*, (6), 81–94.
Evans, L. 1991. *Traffic Safety and the Driver.* New York: Van Nostrand Reinhold.
Evans, L. 1994. Cycle helmets and the law: Even when the science is clear policy decisions may still be difficult, *British Medical Journal*, (308), 1521–2.
FEMA. 2004. *European Agenda for Motorcycle Safety.* Brussels: The Federation of European Motorcyclists' Associations (FEMA).
Fessler, D.M.T., Pillsworth, E.G and Flamson, T.J. 2004. Angry men and disgusted women: An evolutionary approach to the influence of emotions on risk taking. *Organizational Behavior and Human Decision Process,* 95, 107–23.
Festinger, L. 1950. Informal social communication. *Psychological Review*, 57, 271–82.
Festinger, L .1953. An analysis of compliant behaviour, in *Group Relations at the Crossroads,* edited by M. Sherif and M.O. Wilson. New York: Harper and Row.
Fischer, S. and Smith, G.T. 2004. Deliberation affects risk taking beyond sensation seeking. *Personality and Individual Differences*, 36(3), 527–37.
Flemming, H.S. and Becker, E.R. 1992. The impact of the Texas 1989 motorcycle helmet law on total and head-related fatalities, severe injuries, and overall injuries. *Medical Care*, 30, 832–45.

Ford. 2007. *Geneva Global Debut for New Fiesta*. [Online]. Available at: http://www.ford.co.uk/ns7/vehicle_news/-/vhcl_news_article91 [accessed: 30 November 2007].

Freixanet, M.G. 1991. Personality profile of subjects engaged in high physical risk sports. *Personality and Individual Differences*, (12), 1087–93.

Freudenburg, W.R. 1998. Perceived risk, real risk: Social science and the art of probabilistic risk assessment. *Science*, 242(4875), 44–9.

Fuller, R. 2005. Towards a general theory of driver behaviour. *Accident Analysis and Prevention*, 37(3), 461–72.

Fuller, R. 2008. Driver training and assessment: Implications of the task-difficulty homeostasis model, in *Driver Behaviour and Training, Vol 3*, edited by L. Dorn. Aldershot: Ashgate Publishing.

Fuller, R., Bates, H., Gormley, M., Hannigan, B., Stradling, S., Broughton, P.S., Kinnear, N. and O'Dolan, C. 2006. Inappropriate high speed: Who does it and why?, in *Behavioural Research in Road Safety 2006, Sixteenth Seminar*. London: DfT.

Gifford, R. 2003. *Despite Cameras and Promises, Road Deaths Rise to 3,500 a Year.* [Online]. Available at: http://www.timesonline.co.uk/tol/news/uk/article449751.ece [accessed: 7 October 2008].

Gigerenzer, G. and Todd, P.M. 1999. *Simple Heuristics That Make Us Smart*. Oxford: Oxford University Press.

Gill, D.L. 1986 *Psychological Dynamics of Sport and Exercise*. Leeds: Human Kinetics.

Global Positioning Systems. 2007. *Motorbike Systems*. [Online]. Available at: http://www.globalpositioningsystems.co.uk/c-motorbike-apl.html [accessed: 13 February 2007].

Goldenbeld, C., Twisk, D. and Craen, S. 2004. Short and long term effects of moped rider training: A field experiment. *Transportation Research Part F: Traffic Psychology and Behaviour*, 7(1), 1–16.

Graf, P. and Schacter, D.L. 1985. Implicit and explicit memory for new associations in normal and amnesic subjects. *Journal of Experimental Psychology: Learning, Memory, and Cognition*, 11, 501–18.

Grayson G.B. 1996. *Behaviour Ddaptation: A Review of the Literature*. Crowthorne: TRL Limited.

Grayson, G.B. Maycock, G., Groeger, J.A., Hammond, S.M. and Field, D.T. 2003. *Risk, Hazard Perception and Perceived Control.* Crowthorne: TRL Limited.

Griffin, J. 2002. *Well-being: Its Meaning, Measurement and Moral Importance.* Oxford: Clarendon Press.

Guardian. 1999. *Riders on the Storm*. [Online]. Available at: http://www.guardian.co.uk/theguardian/1999/feb/13/weekend7.weekend5 [accessed: 12 December 2005].

Gutkind, L. 2008. *Bike Fever.* River Grove: Follett.

Haberlandt, K. 1999. *Human Memory: Exploration and Application.* Maryland: Allyn & Bacon.

Haigney, D.E., Taylor, R.G. and Westerman, S.J. 2000. Concurrent mobile (cellular) phone use and driving performance: Task demand characteristics and compensatory processes. *Transportation Research Part F: Traffic Psychology and Behaviour*, 3, 113–21.

Hall, S. 1989. The meaning of new times, in *The Changing Face of Politics in the 1990s*, edited by S. Hall and M. Jacques. London: Lawerence and Wishart.

Hannigan, B., Fuller, R., Bates, H., Gormley, M., Stradling, S., Broughton, P.S., Kinnear, N. and O'Dolan, C. 2008. Understanding inappropriate high speeds by motorcyclists: A qualitative analysis, in *Driver Behaviour and Training, Vol 3,* edited by L. Dorn. Aldershot: Ashgate Publishing.

Happian-Smith, J. and Chinn, B.P. 1990. *Simulation of Airbag Restraint Systems in Forward Impacts of Motorcycles.* Detroit: International Congress and Exposition.

Harre, N. 2000. Risk evaluation, driving, and adolescents: A typology. *Developmental Review*, 20(2), 206–26.

HarrowDrive. 2007. *Getting the Right Attitude-Self Talk.* [Online]. Available at: http://www.harrowdrive.com/?p=107 [accessed: 19 July 2007].

Hatakka, M., Keskinen, E., Gregersen, N.P., Glad, A.and Hernetkoski, K. 2002. From control of the vehicle to personal self-control; broadening the perspectives to driver education. *Transportation Research Part F: Traffic Psychology and Behaviour*, 5(3), 201–15.

Hatzigeorgiadis, A. 2002. Thoughts of escape during competition: relationships with goal orientations and self-consciousness. *Psychology of Sport and Exercise*, 3(3), 195–207.

Haworth, N. and Rowden, P. 2006. Fatigue in motorcycle crashes. Is there an issue?, in *Proceedings of Queensland. Australasian Road Safety Research, Policing and Education Conference.* Queensland: Australasian Road Safety Research.

Heino, A., van der Molen, H.H. and Wilde G.J.S. 1996. Risk perception, risk taking, accident involvement and the need for stimulation. *Safety Science*, 22(1–3), 35–48.

Hells Satans. 2008. *The Hell's Satans*. [Online]. Available at: http://hells-satans.com/ [accessed: 22 July 2008].

Hennie, J. 2008. *Our View on Helmet Laws: Motorcycle madness.* [Online]. Available at: http://blogs.usatoday.com/oped/2008/04/our-view-on-hel.html [accessed: 22 July 2008].

Hewson, C., Yule, P., Laurent, D.and Vogel, C. 2003. *Internet Research Methods*. London: Sage.

Hills, P., Argyle, M. and Reeves, R. 2000. Individual differences in leisure satisfactions: An investigation of four theories of leisure motivation. *Personality and Individual Differences*, 28(4), 763–79.

HM Revenue and Customs. 2002. *HMRC Reference: Notice 701/23 (March 2002).* [Online]. Available at: http://customs.hmrc.gov.uk/channelsPortalWebApp/channelsPortalWebApp.portal?_nfpb=true&_pageLabel=pageLibrary_PublicNoticesAndInfoSheets&propertyType=document&columns=1&id=HMCE_CL_000107 [accessed: 15 June 2008].

Hogarth, R.M. 2001. *Educating Intuition.* Chicago: The University of Chicago Press.

Hogg, M.A and Abrams, D. 1988. *Social Identifications: A Social Psychology of Intergroup Relations and Group Processes*. London: Routledge.

Holt, D.B. 2004. *How Brands Become Icons.* New York: Harvard Business Press.

Hopwood, B. 1998. *Whatever Happened to the British Motorcycle Industry?* Yeovil, Somerset: Haynes Publishing.

Horswill, M.S. and McKenna, F.P. 2004, Drivers' hazard perception ability: Situation awareness on the road, in *A Cognitive Approach to Situation Awareness: Theory, Measurement and Application*, edited by S. Branbury and S. Tremblay. Aldershot: Ashgate Publishing.

Horvath, P. and Zuckerman, M. 1993. Sensation seeking, risk appraisal, and risky behaviour. *Personality and Individual Differences*, 1, 41–52.

Huang, B. and Preston, J. 2004. *A Literature Review on Motorcycle Collisions Final Report.* Oxford: Transport Studies Unit Oxford University.

Hurt, H.H., Ouellet, J.V. and Thom, D.R. 1981. *Motorcycle Accident Cause Factors and Identification of Countermeasures. Volume 1: Technical Report.* Washington DC: US Department of Transportation, National Highway Traffic Safety Administration.

Hutchingson, T.P. 2008. Concerns about methodology used in real-world experiments on transport and transport safety. *Journal of Transport Engineering*, Jan 2007, (1), 30–8.

Institute of Highway Incorporated Engineers. 2005. *IHIE Guidelines for Motorcycling. 2005.* Essex: IHIE.

Insurance Institute for Highway Safety. 2008. *Helmet Use Laws.* [Online]. Available at: http://www.iihs.org/laws/HelmetUseOverview.aspx [accessed: 15 June 2008].

Iversen, H. 2004. Risk-taking attitudes and risky driving behaviour. *Transportation Research Part F: Traffic Psychology and Behaviour*, 7(3), 135–50.

Jordon, C., Jordon, C., Hetherington, O., Woodside, A. and Harvey, H. 2004. Noise Induced Hearing Loss in Occupational Motorcyclists, *Journal of Environmental Health Research*, 3(2), 66–73.

Kahane, C.J. 1996. *Fatality Reduction by Air Bags: Analyses of Accident Data Through Early 1996. DOT HS808470.* Washington DC: US Department of Transportation.

Kammann, R. 1983. Objective circumstances, life satisfactions, and sense of well-being: Consistencies across time and place. *New Zealand Journal of Psychology*, 12, 14–22.

Keys, H. 2006. *Biker Tourism: Attracting Motorcyclists to NI.* [Online]. Available at: http://www.bikerroadtrip.com/Manual.pdf [accessed: 8 October 2008].

Kibele, A. 2006. Non-consciously controlled decision making for fast motor reactions in sports – A priming approach for motor responses to non-consciously perceived movement features. *Psychology of Sport and ExerciseJudgement and Decision Making in Sport and Exercise*, 7(6), 591–610.

Kinnear, N., Stradling, S. and McVey, C. 2008. Do we really drive by the seat of our pants, in *Driver Behaviour and Training, Vol 3*, edited by L. Dorn. Aldershot: Ashgate Publishing.

Klein, G. 1998. *Sources of Power: How People Make Decisions*. Cambridge MA: MIT.

Kraus, J.F., Peek, C., McArthur, D.L. and Williams, A. 1994. The effect of the 1992 California motorcycle helmet use law on motorcycle crash fatalities and injuries. *JAMA: The Journal Of The American Medical Association*, 272(19), 1506–11.

Laberge-Nadeau, C., Maag, U., Bellavance, F., Lapierre, S.D., Desjardins, D., Messier and Saidi, A. 2003. Wireless telephones and the risk of road crashes. *Accident Analysis and Prevention*, 35, 649–60.

Lamble, D., Kauranen, T., Laakso, M. and Summala, H. 1999. Cognitive load and detection thresholds in car following situations: Safety implications for using mobile (cellular) telephones while driving. *Accident Analysis and Prevention*, 31, 617–23.

Lancashire Evening Telegraph. 2004. *Shot Fired by Biker Robbers*. [Online]. Available at: http://archive.thisislancashire.co.uk/2004/8/16/473477.html [accessed: 16 August 2006].

Langley, J., Mullin, B., Jackson, R. and Norton, R. 2000. Motorcycle engine size and risk of moderate to fatal injury from a motorcycle crash. *Accident Analysis and Prevention*, 32(5), 659–63.

Lawrence Livermore National Laboratory. 2006. *Environmental, Safety and Health Manual 2006.* California: University of California.

Lehoten, T. and Maenpaa, P. 1997. Shopping in east center mall, in *The Shopping Experience,* edited by P. Falk and C. Campbell. London: Sage.

Leitner, M.J. and Sara, F.L. 2004. *Leisure Enhancement*. New York: The Haworth Press.

Lewin, K. 1935. *A Dynamic Theory of Personality*. New York: McGraw-Hill.

Lewin, K. 1965. Group decision and social change, in *Basic Studies in Social Psychology,* edited by H. Proshansky and B. Seidenberg. New York: Holt, Rinehart and Winston.

Lupton, D. 1999. *Risk.* New York: Routledge.

Lyubomirsky, S., Schkade, D. and Sheldon, K.M. 2005. Pursuing happiness: The architecture of sustainable change. *Review of General Psychology*, 9(2), 111–31.

Mackenzie, B. 2007. *Psychology*. [Online]. Available at: http://www.brianmac.demon.co.uk/psych.htm [accessed: 5 January 2008].

MAG. 2008. *The Heart and Soul of Biking*. [Online]. Available at: http://www.mag-uk.org/ [accessed: 15 October 2008].

Magazzu, D., Comelli, M. and Marinoni, A. 2006. Are car drivers holding a motorcycle licence less responsible for motorcycle-car crash occurence? A non-parametric approach. *Accident Analysis and Prevention*, (38), 365–70.

Magner, T. 2006. *BMF Warn on Driving Licence Threat.* [Online]. Available at: http://www.politics.co.uk/press-releases/public-services/road/road/bmf-warn-on-driving-licence-threat-$459995.htm [accessed: 5 March 2008].

Mannering, F.L. and Grodsky L.L. 1995. Statistical analysis of motorcyclists' perceived accident risk. *Accident Analysis and Prevention*, 27(1), 21–31.

Mano, H. and Elliott, M. 1997. Smart shopping: the origins and consequences of price savings, in *Advances in Consumer Research,* edited by D. MacInnis and M. Brucks. Duluth: Association for Consumer Research.

Massimini, F. and Massimo, C. 1988. The systematic assessment of flow in daily experience, in *Optimal Experience: Psychological Studies of Flow in Consciousness,* edited by M. Csikszentmihalyi and I. Csikszentmihalyi. New York: Cambridge University Press.

Mayou, R. and Bridget, B. 2003. Consequences of road traffic accidents for different types of road user. *Injury*, 34(3), 197–202.

Mazda. 2007. *Geneva Motor Show 2007.* [Online]. Available at: http://www.mazda.co.uk/AboutMazda/MazdaNews/LatestHeadlines/Articles/news_06-Mar-2007 [accessed: 28 March 2007].

McDonald-Walker, S. 2000. *Bikers, Culture, Politics and Power.* Oxford: Berg.

McGwin, G. Jr., Whatley, J., Metzger, J., Valent, F., Barbone, F. and Rue, L.W. 3rd edn. 2004. The effect of state motorcycle licensing laws on motorcycle driver mortality rates. *J Trauma*, 56(2), 415–19.

MCIA. 2006a. *A Street, a Track, an Open Road.* [Online]. Available at: http://www.stordvd.com/ [accessed: 14 March 2007].

MCIA. 2006b. *Research Findings Reveal Post-test Training Valued by Majority if Riders*. [Online]. Available at: http://www.mcia.co.uk/_Attachments/531_101CMS.pdf [accessed: 21 April 2007].

MCIA. 2007a. *About the MCIA.* [Online]. Available at: http://www.mcia.co.uk/S%5FPublic/ [accessed: 21 April 2007].

MCIA. 2007b. *December 2006 New Registration Figures for Motorcycles, Mopeds and Scooters*. Coventry: MCIA.

MCIA. 2007c. *Third European Driving Licence Directive – What You Can Do To Help*. [Online]. Available at: http://www.mcia.co.uk/S_Press/Wizard.asp [accessed: 21 November 2007].

MCIA. 2008a. *International Motorcycle and Scooter Show*. [Online]. Available at: http://www.motorcycleshow.co.uk/ [accessed: 10 October 2008].

MCIA. 2008b. *National Motorcycle Week and Ride to Work Day 25th – 31st July.* [Online]. Available at: http://www.mcia.co.uk/_Attachments/140_101CMS.pdf [accessed: 20 July 2008].

McKenna, F.P. 1985. Do safety measures really work? An examination of risk homoeostasis theory. *Ergomonics*, 28, 489–98.

McKenna, F.P. 1988. What role should the concept of risk play in theories of accident involvement? *Ergomonics*, 31, 469–84.

McKnight, J. and McKnight, S. 1995. The effects of motorcycle helmets upon seeing and hearing. *Accident Analysis and Prevention*, 27(4), 493–501.

Milgram, S. 1964. Group pressure and action against a person. *Journal of Abnormal Social Psychology*, 69, 137–43.

Miller, W.L. and Crabtree, B.F. 1992. Primary care research: A multimethod typology and qualitative road map, in *Doing Qualitative Research*, edited by W.L. Miller and B.F. Crabtree. Newbury Park: Sage.

Mills N.J. 1996. Accident investigation of motorcycle helmets. *Impact (Journal of the institute of traffic accident investigators)*, Autumn (5), 46–51.

Mintel. 2004. *Motorcycles and Scooters – UK – April 2004.* London: Mintel International Group Limited.

Morris, B. 1987. As a favored pastime, shopping ranks high with most Americans. *Wall Street Journal*, 1, 13.

Motorcycle Action Group. 2005. *Vehicle Restraint Systems: Safety Fences: Crash Barriers: Motorcyclists.* Rugby: MAG.

Motorcycle UK Ltd. 2007. *BikeSafe*. [Online]. Available at: http://www.bikesafe.co.uk/ [accessed: November 2007].

Motorcycle.org.uk. 2008. *Back Patches, Top Rockers, Bottom Rockers*. [Online]. Available at: http://www.motorcycle.org.uk/ [accessed: 19 September 2008].

Muelleman, R.L., E.J. Mlinek, E.J. and Collicott, P.E. 1992. Motorcycle crash injuries and costs: Effect of a reenacted comprehensive helmet use law. *Annals of Emergency Medicine*, 21, 266–72.

Näätänen, R. and Summala, H. 1976. *Road User Behaviour and Traffic Accidents. Amsterdam.* New York: North Holland/Elsevier.

NABD. 2008a. *Adaptations*. [Online]. Available at: http://www.nabd.org.uk/adaptions/javaindex.htm [accessed: 15 June 2008].

NABD. 2008b. *NABD by the Trossachs*. [Online]. Available at: http://www.nabd.org.uk/events/2008/August/NABD_By_the_Trossachs_2008-564.html [accessed: 14 July 2008].

Nasctmento Silva, P. 2007. *Review of Vehicle Excise Duty Evasion Statistics.* Southampton: Southampton Statistical Science Research Institute.

National Motorcycle Council. 2000. NMC Policy. [Online]. Available at: http://www.despatch.co.uk/nmc/policy.htm [accessed: 2 February 2007].

Noland, R.B. 1994. Policy simulations of commuter responses to travel time uncertainty under congested conditions. *Paper presented at the Western Regional Science Association Meeting*, October 1994.

NRS. 2006. *Demographics Classifications.* [Online]. Available at: http://www.businessballs.com/demographicsclassifications.htm [accessed: 4 February 2007].

O'Neill, B. and Lund, A.K. 1992. The effectiveness of air bags in preventing driver fatalities in the United States. *International Conference on Air Bags and Seat Belts: Evaluation and Implications for Public Policy* (Montreal, Canada, October 18–20, 1992).

O'Neill, B. and Williams, A. 1998. Risk homeostasis hypothesis: A rebuttal. *Injury Prevention*, (4), 92–3.

Office for National Statistics. 2004. *Family Spending – A Report on the Expenditure and Food Survey*. London: Office for National Statistics.

Orit Taubman, B., Mikulincer, M. and Iram, A. 2004. A multi-factorial framework for understanding reckless driving-appraisal indicators and perceived environmental determinants. *Transport Research.Part F*, 7, 333–49.

Ormston, R., Dudleston, A., Pearson, S. and Stradling, S. 2003. *Evaluation of Bikesafe Scotland*. Edinburgh: Scottish Executive.

Owl Research Ltd. 2008. *Ride Course*. [Online]. Available at: http://www.ride-course.org/index.htm [accessed: 25 July 2008].

Oxford Concise Dictionary. *Pleasure*. 2001. Oxford: Oxford University Press.

O'Sullivan, D.M., Zuckerman, M. and Kraft, M. 1998. Personality characteristics of male and female participants in team sports. *Personality and Individual Differences*, 25, 119–28.

Panou, M., Bekiaris, E. and Papakostopoulos, V. 2005. Modeling driver behaviour, in *EU and International Projects*, edited by L. Macchi and P.C. Cacciabue. Ispra, Luxembourg: Office for Official Publication of the European Communities.

Pao, Y. 1989. *Adaptive Pattern Recognition and Neural Networks*. New York: Addison-Wesley.

Peltzeman, S. 1975. The effects of automobile safety regulations. *Journal of Political Economy*, 83, 677–725.

Petterson, S., Hoffer, G. and Millner, E. 1995. Are drivers of air-bag equipped cars more aggressive? A test of the offsetting behavior hypothesis. *Journal of Leisure Research*, 38, 251–64.

Preusser, D.F., Williams, A.F. and Ulmer, U.G. 1995. Analysis of fatal motorcycle crashes: Crash typing. *Accident Analysis and Prevention*, 27(6).

Prochaska, J.O., DiClemente, C.C. and Norcross, J.C. 1992. In search of how people change-applications to addictive behaviors. *American Psychologist*, 47(9), 1102–114.

Quiñones, L. 2006. *Rebel Image of Motorcyclists Set in 1950s*. [Online]. Available at: http://www.epcc.edu/nwlibrary/borderlands/12_rebel_image_of_motorcyclist.htm [accessed: 25 May 2006].

Raab, M. 2003. Decision making in sports: Implicit and explicit learning is affected by complexity of situation. *International Journal of Sport and Exercise Psychology*, 1, 406–33.

Rohrmann, B. 2002. *Risk Attitude Scales: Concepts and Questionnaires*. Melbourne: University of Melbourne.

Rooney, F. 1951. Cyclists' raid. *Harper's Magazine*. January 1951, 34–44.

RoSPA. 2001. *Motorcycling Safety Position Paper.* Birmingham: The Royal Society for the Prevention of Accidents.

RoSPA. 2006. *Ride Safe*. [Online]. Available at: http://www.rospa.com/roadsafety/ridesafe/ [accessed: 10 October 2008].

RoSPA. 2007. *Advanced Motorcycle Training: Advanced Tests*. [Online]. Available at: http://www.rospa.com/drivertraining/courses/advanced_tests/motorcycle_scooter.htm [accessed: 28 April 2008].

RoSPA. 2008. *Advanced Tests: Car*. [Online]. Available at: http://www.rospa.com/drivertraining/courses/advanced_tests/car.htm . 2008 [accessed: 28 April 2008].

Ross, D. 2008. *Motorcycle Test Centres to be Reduced from 64 to 12*. [Online]. Available at: http://www.theherald.co.uk/news/news/display.var.2193514.0.Motorcycle_test_centres_to_be_reduced_from_64_to_12.php [accessed: 10 October 2008].

Rothe, J.P. and Cooper, P.J. 1987. *Motorcyclists: Image and Reality*. Canada: Insurance Corporation of British Columbia.

Rothschild, Lord. 1979. Coming to grips with risk. *The Wall Street Journal*, 13 March 1979, 22.

Routledge, P. 2008. *Bikers are Jammy Dodgers*. [Online]. Available at: http://www.mirror.co.uk/news/columnists/routledge/2008/01/25/bikers-are-jammy-dodgers-89520-20297209/ [accessed: 25 January 2008].

Rutter, D.R. and Quine. L. 1996. Age and experience in motorcycling safety. *Accident Analysis and Prevention*, 28(1), 15–21.

Scanlan, T. K. and Lewthwaite, R. 1984. Social psychological aspects of competition for male youth sport participants: I. Predictors of competitive stress. *Journal of Sport Psychology*, 6, 208–26.

Scanlan, T.K. and Simons. J.P. 1992. The construct of sport enjoyment, in *Motivation in Sport and Exercise*, edited by G.C. Roberts. Champaign, IL: Human Kinetics Publishing.

Scanlan, T.K., Stein, G.L. and Ravizza, K. 1991. An in-depth study of former elite figure skaters: III. Sources of stress. *Journal of Sport and Exercise Psychology*, 13, 103–20.

Schindler, R. 1989. The excitement of getting a bargain: some hypotheses concerning the origins and effects of smart-shopper feelings, in *Advances in Consumer Research*, edited by T. Srull. New York: Association for Consumer Research.

ScotlandByBike. 2008. *Self Guided Tours*. [Online]. Available at: http://www.scotlandbybike.com/selfguidedtours.asp [accessed: 25 August 2008].

Scottish Executive. 2006. *Key 2005 Road Accident Statistics.* Edinburgh: Scottish Executive.

Segerstorm, S.C. 2006. *Breaking Murphy's Law: How Optimists Get What They Want from Life-and Pessimists Can Too.* Guilford: Guilford Press.

Sexton, B., Baughan, C.J., Elliot, M.A. and Maycock, G. 2004. *The Accident Risk of Motorcyclists. TRL607.* Crowthorne: Transport Research Laboratory.

Sexton, B., Fletcher, J. and Hamilton, K. 2004. *Motorcycle Accidents and Casualties in Scotland 1992-2002*. Edinburgh: Scottish Executive.

Sexton, H., Baughan, C.J., Stradling, S. and Broughton, P.S. 2006. *Risk and Motorcyclists in Scotland*. Edinburgh: Scottish Executive.

Shaoul, J. 1975. *The Use of Accidents and Traffic Offenses as Criteria for Evaluating Courses in Driver Education*. Manchester: The University of Salford.

Sharp, B. 2001. *Strategies for Improving Mountain Safety*. Glasgow: Strathclyde University/Leverhulme Trust.

Sher, J. and Marsden, W. 2006. *Angels of Death: Inside the Bikers' Global Crime Empire*. New York: Da Capo Press.

Siegrist, M., Cventkovich, G. and Gutscher, H. 2002. Risk preference predictions and gender stereotypes. *Organizational Behaviour and Human Decision Processes*, 87, 233–52.

Slanger, E. and Rudestam, K.E. 1997. Motivation and disinhibition in high risk sports: Sensation seeking and self-efficacy. *Journal of Research in Personality*, 31(3), 355–74.

Smith, C. 2007. *Congestion Charge: Potentially Unsafe for Motorcyclists, Claim Researchers*. [Online]. Available at: http://www3.imperial.ac.uk/newsandeventspggrp/imperialcollege/newssummary/news_17-8-2007-10-27-41 [accessed: 12 October 2007].

Smith, R.E. and Smoll, F.L. 1977. Coaching the coaches: Youth sports as a scientific and applied behavioural setting. *Current Directions in Psychology Science*, 6, 16–21.

South Gloucestershire Council. 2008. *The 'Sorry Mate I Didn't See You' Campaign*. [Online]. Available at: http://www.smidsy.co.uk/ [accessed: 9 May 2008].

Sporner, A. and Kramlich, T. 2000. *Zusammenspiel von aktiver und passiver Sicherheit bei Motorradkollisionen*. München: Intermot 2000.

Sporner, A., Langwieder, K., and Polauke. 1990. *Passive Safety for Motorcyclists – from the Legprotector to the Airbag*. Detroit: International Congress and Exposition.

Staffordshire County Council. 2005. *Local Transport Plan 2006–2011*. Stafford: Staffordshire County Council.

Steg, L. 2004. Instrumental, social and affective values of car use. In *Proceedings of 3rd International Conference on Traffic & Transport Psychology*.

Steg, L., Vlek, C. and Slotegraaf, A. 2001. Instrumental-reasoned and symbolic-affective motives for using a motor car. *Transportation Research Part F: Traffic Psychology and Behaviour*, 4, 151–69.

Stone, J. 2008. *BMF Welcomes Apology Over Motorcycle Tax Evasion*. [Online]. Available at: http://www.politics.co.uk/press-releases/bmf-welcomes-apology-over-motorcycle-tax-evasion-$1208412.htm [accessed: 28 February 2008].

Stradling, S.G. 2003. Reducing car dependence, in *Integrated Futures and Transport Choices*, edited by J. Preston. Aldershot: Ashgate Publishing.

Stradling, S.G. 2005. *At Risk on the Road – Why Young Drivers Die*. London: Foundation for the Automobile and Society.

Stradling, S.G. and Anable, J. 2007. Individual Travel Patterns, in *Transport Geographies: An Introduction,* edited by R.D. Knowles, J. Shaw, and I. Docherty. Oxford: Blackwell Publishers.

Stradling, S.G. and Cunliffe, N. 2006. *Riders Improving and Developing Their Experience: RiDE Programme*. Blackpool: RoSPA.

Stradling, S.G., Meadows, M.L. and Beatty, S. 2001. Identity and independence: Two dimensions of driver autonomy, in *Behavioural Research in Road Safety XI,* edited by G. Grayson. Crowthorne: Transport Research Laboratory.

Stuart, J. 1987. *Rockers!* London: Plexus Publishing Ltd.

Sudlow, D. 2003. *Road Safety Research Report No 36: Scoping Study on Motorcycle Training*. London: British Institute of Traffic Education Research.

Summala, H. 1986. *Risk Control is Not Risk Adjustment: The Zero-risk Theory of Driver Behavior and its Implications. Report 11*. Helsinki: University of Helsinki Traffic Research Unit.

Tauber, E. 1972. Why do people shop. *Journal of Marketing*, 36, 46–59.

Taylor, D.H. 1964. Drivers' galvanic skin response and the risk of accident. *Ergomonics*, 7, 439–51.

Telegraph Media Group. 2007. *BMW R1200GS*. [Online]. Available at: http://www.telegraph.co.uk/core/Slideshow/slideshowContentFrameFragXL.jhtml?xml=/motoring/galleries/motorcycles/bmw/2007/k1200gs/pixbmw.xml&site=arts [accessed: 15 June 2008].

Tenenbaum, G. and Lidor, R. 2005. Research on decision-making and the use of cognitive strategies in sport settings, in *Handbook of Research in Applied Sport Psychology: International Perspectives,* edited by D. Hackfort, J.L. Duda, and R. Lidor. Morgantown, West Virginia: Fitness Information Technology.

The Association of Chief Police Officers. 2003. *National Driver Improvement Scheme: Guidance Notes*. London: ACPO.

The Countryside Agency. 2002. *Two Wheels Work.* Wetherby: Countryside Agency Publications.

The Crown Prosecution Service. 2006. *Road Traffic Offences*. [Online]. Available at: http://www.cps.gov.uk/legal/section9/chapter_a.html [accessed: 15 July 2007].

The Hairy Bikers. 2008. *The Hairy Bikers*. [Online]. Available at: http://www.hairybikers.com/blog/ [accessed: 18 June 2008].

The RAC Foundation. 2002. *Motoring Towards 2050 – The Cleaner, Leaner. Safer Car*. [Online]. Available at: http://www.racfoundation.org/index.php?option=com_content&task=view&id=236&Itemid=0 [accessed: 6 March 2005].

The Scottish Motorcycle Club. 2006. *The Scottish Motorcycle Club.* [Online]. Available at: http://www.scottishmotorcycleclub.org.uk/ [accessed: 25 January 2007].

Thompson, C.E and Wankel, L.M. 1980. The effects of perceived activity choice upon frequency of exersize behaviour. *Journal of Applied Social Psychology*, 10, 91–9.

Thompson, D.C., Thompson, R.S. and Rivara F.P. 2001. Risk compensation theory should be subject to systematic reviews of the scientific evidence, *Injury Prevention*, 7, 86–8.

Thorndike, E.L. and Rock, R.T. 1934. Learning without awareness of what is being learned or intent to learn it. *Journal of Experimental Psychology*, 17, 1–19.

Time Magazine. 2008 *The Wilder Ones.* [Online]. Available at: http://www.time.com/time/magazine/article/0,9171,841749,00.html [accessed: 1 June 2008].

Torkildsen, G. 2005. *Leisure and Recreation Management*. London: Routledge.

Townsend, M. 2006. Motorists' use of hand held cell phones in New Zealand: An observational study. *Accident Analysis and Prevention*, 38(4), 748–50.

Turner, C. and McClure, R. 2004. Quantifying the role of risk-taking behaviour in causation of serious road crash-related injury. *Accident Analysis and Prevention*, 36(3), 383–9.

Turner, C., McClure, R. and Pirozzo, S. 2004. Injury and risk-taking behavior – A systematic review. *Accident Analysis and Prevention*, 36(1), 93–101.

Turner, J.C. 1991. *Social Influences.* Milton Keynes: Open University Press.

USA Today. 2008. *Our View on Helmet Laws: Motorcycle Madness*. [Online]. Available at: http://blogs.usatoday.com/oped/2008/04/our-view-on-hel.html [accessed: 2 June 2008].

Vaca, F. 2006. Commentary: Motorcycle helmet law repeal: When will we learn. *Annals of Emergency Medicine*, 47, 204–6.

Van Boven, L. 2005. Experientialism, materialism, and the pursuit of happiness. *Review of General Psychology*, 9(2), 132–42.

Veno, A. 2003. *The Brotherhoods: Inside the Outlaw Motorcycle Clubs*. Saint Leonards, Australia: Allen and Unwin.

VicRoads. 2001. *Motorcycling Notes – Designing for Motorcycle Clearances*. Victoria, Australia: VicRoads.

VisitScotland. 2006. *Bikers Welcome Scheme.* [Online]. Available at: http://www.visitscotland.org/bikers_criteria_-_serviced___self-catering.pdf [accessed: 2 September 2008].

Walker, L. 2007. *Scoping the Dimensions of Visitor Well-being: A Case Study of Scotland's Forth Valley*. Unpublished PhD thesis, Department of Marketing, University of Stirling.

Wallace, P., Haworth, N. and Regan, M. 2005. *Best Training Methods for Teaching Hazard Perception and Responding by Motorcyclists*. Monash, Australia: Monash University Accident Research Centre.

Wang, J. 2002. *Developing and Testing an Integrated Model of Choking in Sport*. Unpublished PhD thesis, Victoria University.

Wankel, L.M. and Kreisel, P.S. 1985. Factors underlying enjoyment of youth sports: sport and age group comparisons. *Journal of Sport Psychology*, 1, 51–64.

Watson, G.S., Zador, P.L. and Wilks, A. 1981. Helmet use, helmet use laws, and motorcyclist fatalities. *American Journal of Public Health*, 71, 297–300.

Webster's 1979. *Webster's New World Dictionary of the American Language*. New York: Simon & Schuster.

West Coast Harely Tours. 2008. *West Coast Harely Tours*. [Online]. Available at: http://www.westcoastharleytours.com/ [accessed: 30 August 2008].

White Rose Tours. 2008. *Border Reivers*. [Online]. Available at: http://www.motorcycletours.co.uk/tour13.html [accessed: 30 August 2008].

Wickens, C.D. and Hollands, J.G. 2000. *Engineering Psychology and Human Performance*. New Jersey: Prentice Hall.

Wilde, G.J.S. 1982. The theory of risk homeostasis: Implications for safety and health. *Risk Analysis*, 2, 209–25.

Wilde, G.J.S. 1998. Risk homeostasis theory: An overview. *Injury Prevention*, 4, 89–91.

Williams, J.M. and Leffingwell, T.R. 1996. Cognitive strategies in sport and exercise psychology, in *Exploring Sport and Exercise Psychology,* edited by J.L. Van Raalte and B.W. Brewer. Washington DC: American Psychology Association.

Woollard, D. 2008. The need for speed: A look at motorcycle-loving celebrities. *Luxist*. 14 February 2008.

Wright Mills, C. 1959. *The Sociological Imagination*. New York and Oxford: Oxford University Press.

Wyatt, J.P., O'Donnell, J., Beard, D. and Busuttil, A. 1999. Injury analyses of fatal motorcycle collisions in south-east Scotland. *Forensic Science International*, 104(2-3), 127–32.

Yamaha Motor Company. 2007a. *MT-01 and MT-03: Maximum Fun in Minimum Time.* [Online]. Available at: http://www.yamaha-motor-europe.com/designcafe/en/about/sports_touring/?Component=tcm:71-202474&PageTitle=MT%2003%20-%20Single%20cylinder%20show%20model%20&pageNum=7 [accessed: 25 September 2008].

Yamaha Motor Company. 2007b. *YZF-R1. 2007*. [Online]. Available at: http://www.yamaha-motor.co.uk/products/motorcycles/supersport/yzf_r1.jsp [accessed: 25 September 2008].

Yannis, G., Golias, J. and Papadimitriou, E. 2005. Driver age and vehicle engine size effects on fault and severity in young motorcyclists accidents. *Accident Analysis and Prevention*, 37, 327–33.

Zeigler, B.P. 2002. The brain-machine disanalogy revisited. *Biosystems*, 64(1-3), 127–40.

Zuckerman, M. 1979. *Sensation Seeking: Beyond the Optimal Level of Arousal*. Hillsdale: Earlbaum.

Zuckerman, M. 1983. Sensation seeking and sports. *Personality and Individual Differences*, 4, 285–93.

Zuckerman, M. 1991. Sensation Seeking: The balance between risk and reward, in *Self-regulatory Behaviour and Risk Taking: Causes and Consequences,* edited by L.P. Lipsitt and L.L. Minick. Norwoord: Ablex.

Zuckerman, M. 1992. What is a basic factor and which factors are basic? Turtles all the way down. *Personality and Individual Differences*, (13), 675–81.

Zuckerman, M. 1994. *Behavioral Expressions and Biosocial Bases of Sensation Seeking.* Cambridge: Cambridge University Press.

Index

For Product Safety Concerns and Information please contact our EU
representative GPSR@taylorandfrancis.com
Taylor & Francis Verlag GmbH, Kaufingerstraße 24, 80331 München, Germany

www.ingramcontent.com/pod-product-compliance
Ingram Content Group UK Ltd.
Pitfield, Milton Keynes, MK11 3LW, UK
UKHW021612240425
457818UK00018B/516

* 9 7 8 0 3 6 7 3 8 5 6 0 6 *